# Core Anatomy for Students

**Core Anatomy for Students**

*Other titles also available*

# Core Anatomy for Students

## Volume 2: The Thorax, Abdomen, Pelvis and Perineum

**Christopher Dean**

*Department of Anatomy and Developmental Biology,
University College London,
London, UK*

and

**John Pegington**

*Department of Anatomy and Developmental Biology,
University College London,
London, UK*

WB SAUNDERS COMPANY LTD
London   Philadelphia   Toronto   Sydney   Tokyo

W.B. Saunders Company Ltd    24–28 Oval Road
London NW1 7DX, UK

The Curtis Center
Independence Square West
Philadelphia, PA 19106-3399, USA

Harcourt Brace & Company
55 Horner Avenue
Toronto, Ontario M8Z 4X6, Canada

Harcourt Brace & Company, Australia
30–52 Smidmore Street
Marrickville, NSW 2204, Australia

Harcourt Brace & Company, Japan
Ichibancho Central Building, 22-1 Ichibancho
Chiyoda-ku, Tokyo 102, Japan

British Library Cataloguing in Publication Data is available

ISBN 0-7020-2041-9

This book is printed on acid-free paper

Typeset by Photo·graphics, Honiton, Devon.
Printed and bound by the Bath Press, Avon, UK

*Core Anatomy for Students* is dedicated to the memory of John Pegington. Few people have thought harder about how to teach anatomy for the first time. Few people have been as successful at it.

# Contents

# Preface

*Core Anatomy for Students* was written as a revision text. It was originally intended for students who may have left themselves short of time to work for exams, or for those who may be faced with sitting anatomy exams for a second time. We presume, therefore, that students who use one or all of these three volumes will already have completed an anatomy course and will therefore be familiar with basic anatomical terminology. We expect that they will have studied some developmental biology and histology, and also that those who use this text will be keen to start their clinical studies and will be curious to know why a good deal of anatomy is clinically important. With these things in mind we have occasionally drawn freely on developmental anatomy and some applied anatomy in each of these three volumes to clarify or to illustrate what we feel is important material.

These volumes are not designed to be used as a standard textbook or reference book of anatomy. They are meant to provide a framework for revision and self-directed learning. They represent a synopsis of basic material that we feel defines a core of useful knowledge. We regard this core as material that lies in the current mainstream of a continuous learning programme such as medicine. It is that material which we consider as necessary to know in order to understand the next step in this sequence of learning. The content of each volume is therefore highly selective and many things have deliberately been left out. Neither do we intend *Core Anatomy for Students* to be a set of revision *notes* where factual details are maximally condensed. Our emphasis here is on a readable text that explains, sometimes at length, what may be difficult or important. On occasions the text is deliberately repetitious. We have tried to promote an understanding of anatomy in a way that reinforces the learning process, which we feel is not achieved by lists of things to be revised and committed to memory.

The material presented here is probably not set out in a way that parallels the way it was first taught. We hope that the order in which things are presented makes functional and logical sense, and also ties together some topics that may otherwise seem unrelated. Above all, we hope that each section of each volume forms a cogent revision programme in its own right and that dental, medical, speech science, podiatry, physiotherapy and other students who study anatomy will find *Core Anatomy for Students* useful.

# Introduction

Volume 2 of *Core Anatomy For Students* is set out in three sections. The first section covers the thorax, the second the abdomen and the third the pelvis and perineum. While each of these sections is complete in its own right and can be studied in any order, we suggest the way they are set out may be the most appropriate order to work through them, since occasional reference is made later on to material discussed in the earlier sections.

We have made no attempt to reduce the chapters in each volume to an equal length. As they stand they represent what we feel are coherent functional units which all form part of a sequential revision programme. The illustrations are for the most part designed to be coloured in as you work through the text. We would encourage you at least to choose a few key diagrams in each section to colour in, since there is evidence that this helps to commit the three-dimensional aspects of anatomy to memory. The legends for each diagram can also be used as a summary of the text when revising each section on subsequent occasions.

At the end of each section there is a revision chapter to help consolidate your anatomical knowledge and to test your understanding of each region. You may choose to do the multiple choice questions a few at a time as you finish reading each chapter. You may prefer to do them all together at the end of each section. Alternatively, you might even consider doing the even-numbered questions on your first attempt and the odd-numbered questions on a subsequent occasion. Whatever you decide, remember that they are an integral part of this revision programme and that you need to work through them with reference to the text at some stage to get the most out of your revision. Do not be tempted to ignore them altogether.

# Acknowledgements

In the first instance we would like to acknowledge Dr Wojtek Krzemieniewski who drew earlier versions of some of the illustrations we have used. We are especially grateful to Breda O'Connor for typing a great deal of the manuscript and to Dr Deana D'Souza for her scrutiny of the text and illustrations. We are grateful to Barry Johnson and Derek Dudley for technical support and to Jane Pendjiky and Chris Sym for photographic assistance. In particular, we are grateful to many generations of anatomy, dental, medical, speech science, podiatry and physiotherapy students from The School of Medicine, Ottawa, and from University College London, who have all used earlier versions of this revision text, and who over the years have encouraged us to write more of them, to improve them, correct them and finally to publish them. It goes without saying that the artwork in any anatomy book is fundamental to its success. We are especially grateful to Joanna Cameron for her exceptional illustrations. We would also like to express our thanks to the production team at W.B. Saunders, London.

# THE THORAX

# Introduction to the Peripheral Nervous System

A sound knowledge of the basic organisation of the peripheral nervous system makes anatomy of the thorax, abdomen and pelvis easier to understand. For this reason it is sensible to begin any revision programme on the anatomy of the trunk with an overview of the peripheral nervous system. The finer details of peripheral neuroanatomy can then be built up region by region.

The nervous system is continually synthesizing information about the environment and initiating appropriate reactions to it. The **central nervous system**, or CNS, is housed in the skull and vertebral column. It consists of the brain and spinal cord. It can be referred to as the **neuraxis** (Fig. 1.1). Information is fed into the neuraxis and then, following some level of analysis, instructions flow out. This occurs at many levels up and down the neuraxis in a **segmental manner** (Fig. 1.2). **Nerves**, as we refer to them in topographical anatomy, are bundles of **axons**. Each axon of course also has its own **cell body** containing a nucleus. The entire nerve cell is called a **neuron**. Neurons and their supporting cells that lie partly or wholly outside the central nervous system make up the **peripheral nervous system**. These include, for example, all the cranial and spinal nerves.

Outside the central nervous system collections of cell bodies are found in swellings called **ganglia**. Within the central nervous system they form **nuclei**. Rather confusingly, there are other swellings or ganglia in the peripheral nervous system. These contain

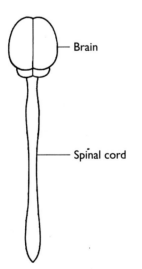

**Figure 1.1** The brain is housed in the skull and the spinal cord in the vertebral column. Together they make up the neuraxis.

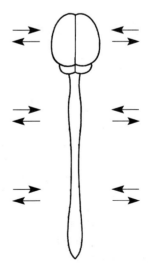

**Figure 1.2** Afferent nerves leave the neuraxis and efferent nerves enter the neuraxis in a segmental manner.

**synapses** between one neuron and another. We need to be clear where each type of ganglion is found.

Some axons are insulated by **Schwann cells** that encircle the axon (Fig. 1.3). These are called **myelinated nerves**. They appear white to the naked eye in fresh material. Other axons are simply embedded in, but not encircled by, their Schwann cells. These axons are called **unmyelinated nerves**. They look grey to the naked eye in fresh material.

**Sensory** or **afferent** nerves form that part of the peripheral nervous system which feeds information *into* the neuraxis. Sensory nerves feed in information about touch, pressure, pain, temperature and position in space, as well as about special senses like smell, taste, hearing and vision. Information *leaves* the neuraxis via **motor** or **efferent** nerves. There are only two types of functioning units in the body innervated by motor nerves: muscles and glands. (If you have difficulty remembering which afferent and efferent nerves are, remember that in effervescence, during a chemical reaction, gas bubbles *out*.)

The peripheral nervous system can be divided into two parts. These are the **somatic nervous system** and the **autonomic nervous system**. The peripheral somatic nervous system acts on 'voluntary structures' such as skeletal muscles, whereas the autonomic nervous system acts on visceral structures such as the salivary glands, the secretory glands of the gastrointestinal tract or the smooth muscle of the heart, gut and bladder.

## The somatic nervous system

Sensory or afferent, and motor or efferent, nerve fibres of the somatic part of the the peripheral nervous system leave or enter the brain and spinal cord in a segmental manner. There are 12 pairs of cranial nerves and 31 pairs of spinal nerves in all.

Sensory spinal nerves form **dorsal nerve roots** that arise from the spinal cord (Fig. 1.4). Sensory nerve fibres have their cells of origin in a ganglion that lies *outside* the the central nervous system (Fig. 1.5). This ganglion is called the **dorsal root ganglion**. Sensory nerves that arise from the brain (cranial nerves) also have their cell bodies outside the central nervous system in a ganglion analogous to a dorsal root ganglion.

Motor nerves leave the spinal cord as **ventral nerve roots** (Fig. 1.4). Unlike sensory nerves they have their

(a)

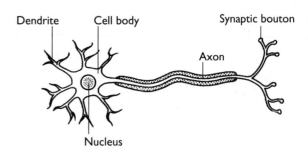

(b)

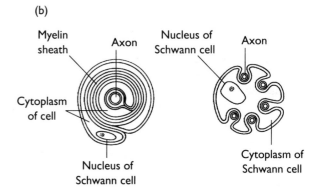

(c)

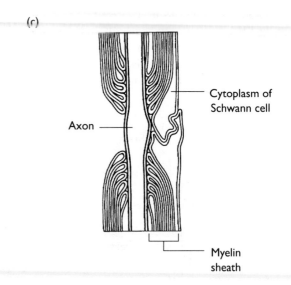

**Figure 1.3** (a) Nerve cells and their components are called neurons. (b) The axons of non-myelinated nerves are sunk into the cytoplasm of their Schwann cells. (b, c) Myelinated nerves have Schwann cells that encircle their axons. (After Aiello LC and Dean MC. *Human Evolutionary Anatomy*. London: Academic Press, 1990.)

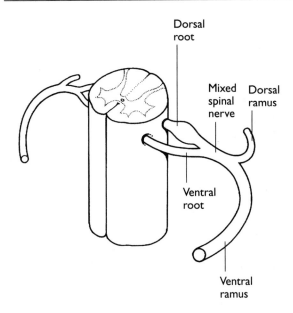

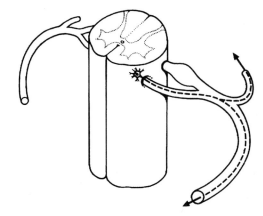

**Figure 1.6**  Motor nerve fibres have their cells of origin within the central nervous system.

**Figure 1.4**  Sensory nerve fibres form dorsal roots and motor nerve fibres form ventral roots close to the spinal cord. These roots join to form a mixed spinal nerve within the intervertebral foramina. The dorsal ramus and the ventral ramus each carry motor and sensory nerve fibres.

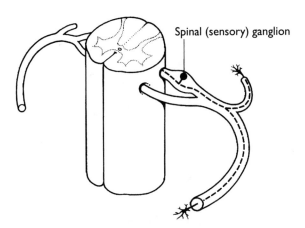

**Figure 1.5**  Sensory nerve fibres have their cell bodies in a spinal or dorsal root ganglion.

cells of origin *within* the central nervous system (Fig. 1.6). In the spinal cord, the cell bodies lie in the **anterior columns** of grey matter. Both motor and sensory spinal nerve roots join up to form a bundle of nerve fibres called a **mixed nerve**. Spinal nerve roots join up to form **mixed spinal nerves** in the region of the intervertebral foramina between two vertebrae. Mixed spinal nerves quickly divide into a **dorsal ramus** and a **ventral ramus**. The dorsal ramus is a bundle of

motor and sensory nerve fibres that supplies the skin and a group of muscles in the midline of the back called the erector spinae muscles. The ventral ramus is a mixture of motor and sensory nerve fibres that supply muscles and skin at the sides and front of the body and the limbs.

Typically, a ventral ramus in the thoracic region runs in an intercostal space and is known as an intercostal nerve here (Fig. 1.7). A **collateral branch** supplies the intercostal muscles with motor fibres in the space. **Lateral cutaneous** branches and **anterior cutaneous** branches pass through to the skin and innervate the side and front of the body respectively. The skin area supplied by a particular nerve is called a **dermatome**. We will say more about mixed spinal nerves in the thorax and anterior abdominal wall in the appropriate sections later on.

Cranial nerves are a bit different because they can be composed completely of motor fibres or of sensory fibres, or they may be mixed nerves like spinal nerves. They can also be nerves concerned with special sensation such as vision, hearing, taste and smell.

## The autonomic nervous system

The autonomic nervous system consists of two motor divisions; the **parasympathetic system** and a **sympathetic system**. Commonly, one system is excitatory and the other inhibitory. For example, the sympathetic nervous system speeds up the heart and dilates the pupil of the eye whereas the parasympathetic nervous system slows down the heart and constricts the pupil. Sometimes, however, visceral structures are supplied

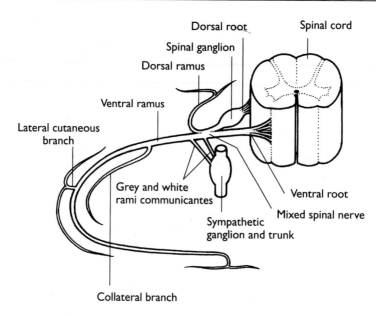

**Figure I.7**   The dorsal ramus supplies motor and sensory nerves to the erector spinae muscles of the back and to a strip of skin in the midline posteriorly. The ventral ramus supplies mixed nerves to the sides and front of the body and the limbs. The ventral ramus of an intercostal space has a collateral branch and both lateral and anterior cutaneous branches.

by only one of the autonomic systems. Smooth muscle in blood vessel walls has only a sympathetic motor nerve supply. The ciliary muscle in the eye, which acts on the lens of the eye for focusing, has only a parasympathetic motor nerve supply. The limbs have no parasympathetic nerve supply at all.

Motor nerves of the autonomic nervous system differ from those of the somatic nervous system in one fundamental way. They have two motor neurons rather than only one in the peripheral part of the nervous system (Fig. 1.8). Autonomic motor nerves therefore **synapse** in a ganglion somewhere between the central nervous system and their target organs. They synapse in **autonomic ganglia**. There must then be a **preganglionic neuron** and a **postganglionic neuron** in both motor sympathetic and motor parasympathetic systems.

A useful general distinction between the sympathetic and parasympathetic motor systems is that sympathetic autonomic ganglia tend to be close to the neuraxis but parasympathetic autonomic ganglia tend to be close to, or even within, their target organs (Fig. 1.9). With this generalization in mind it is now better to study each division of the autonomic nervous system on its own.

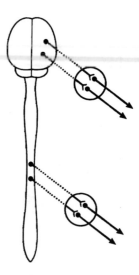

**Figure I.8**   Motor autonomic nerves consist of a preganglionic and a postganglionic neuron that synapse in a ganglion outside the central nervous system.

## The sympathetic nervous system

Sympathetic motor neurons only leave the central nervous system between segmental levels T1 to L2

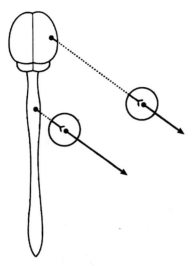

**Figure 1.9** Preganglionic parasympathetic motor neurons (dotted line, in this case running with a cranial nerve from the brain) usually synapse with the postganglionic parasympathetic neurons (solid line) close to their target organs. Preganglionic sympathetic neurons (dotted line) usually synapse with their postganglionic sympathetic neurons (solid line) close to the neuraxis.

inclusive (Fig. 1.10). There has, therefore, to be a way for sympathetic nerves to reach the head and neck, or to reach parts of the body lower than this level such as the abdomen and pelvis. Preganglionic sympathetic nerves have their cell bodies in the lateral column of grey matter in the spinal cord between the levels of T1 and L2. Between the levels of T1 and L2 *only*, they leave the spinal cord in the ventral motor roots along with somatic motor nerves. They then

leave the ventral nerve roots, in myelinated nerves, and pass into a chain of sympathetic nerves that runs either side of the vertebral column from the head to the sacrum (Fig. 1.11). The **sympathetic chain** or **trunk** has ganglia at each segmental level (except, rather confusingly, there are only three in the neck, four in the lumbar and sacral regions, and a single fused one in front of the coccyx because some have coalesced). Each of these ganglia contains synapses between preganglionic and postganglionic sympathetic neurons. Once in the sympathetic trunk, **preganglionic sympathetic neurons** can do one of three things. They can synapse immediately at the segmental level of their origin (Fig. 1.12). They can travel up or down the trunk and synapse at a higher or lower segmental level (Fig. 1.13). They can leave the sympathetic trunk at any level without synapsing at all (Fig. 1.14). Generally, sympathetic nerves synapse at the segmental level closest to the developmental origin of their target organs.

**Postganglionic sympathetic neurons** that have synapsed, either at their own or at a higher or lower segmental level, get to their target organs or glands in

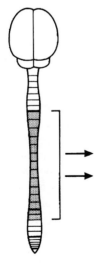

**Figure 1.10** Sympathetic motor neurons only ever leave the neuraxis between levels T1 and L2 of the spinal cord.

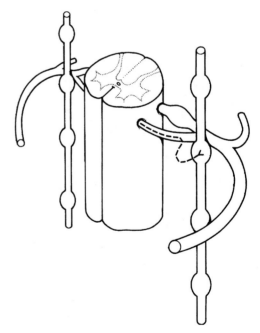

**Figure 1.11** Preganglionic sympathetic nerves (dotted line) leave the spinal cord in the ventral ramus but then pass out of this into the sympathetic trunk. At this point they run on their own and, since they are myelinated and look white in fresh material, they form a white ramus communicans. (Look again at Figure 1.7.)

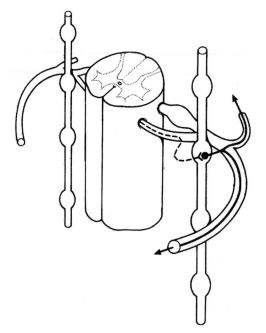

**Figure 1.12**   In this figure a preganglionic sympathetic nerve (dotted line) has synapsed at its own segmental level within the sympathetic trunk. The postganglionic fibres (solid line) are non-myelinated and look grey in fresh material. They run on their own in a grey ramus communicans and then rejoin the mixed spinal nerve.

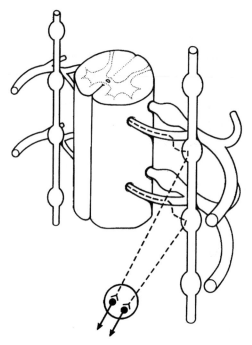

**Figure 1.14**   Some preganglionic sympathetic nerve fibres leave the sympathetic trunk without synapsing and run free (dotted lines) to ganglia some distance from the neuraxis.

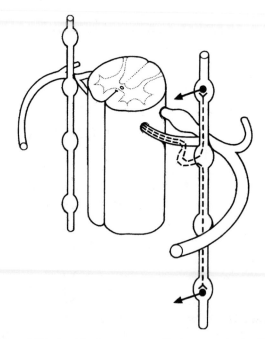

**Figure 1.13**   In this figure preganglionic sympathetic fibres (dotted lines) have travelled further up and down the sympathetic trunk than their own segmental level before synapsing (solid lines).

one of two ways. Some 'jump back', as unmyelinated nerves, into mixed spinal nerves and run along with them to their destination (Fig. 1.15). Others 'jump' on to blood vessels and 'hitch-hike' a ride to their destinations (Fig. 1.16). This happens a lot in the head and neck. But, as we have seen, other postganglionic sympathetic neurons simply go off on their own without synapsing at all in the sympathetic trunk and run as nerves in their own right towards involuntary muscles and glands (Figs 1.14 and 1.17).

Sometimes the myelinated **preganglionic** motor sympathetic nerves that run from the ventral root to the sympathetic trunk are referred to as **white rami communicantes**. Likewise, the unmyelinated **post-ganglionic** sympathetic motor nerves that jump back from the sympathetic trunk into mixed spinal nerves are called **grey rami communicantes** (Fig. 1.7). This is because in living material they look white or grey due to the presence or absence of a thick myelin sheath. Some examples of sympathetic innervation to various visceral structures help to to illustrate what we have just described in a more meaningful way.

The sympathetic supply to the head arises from the spinal cord at the level of T1 and T2. Fibres then run up the sympathetic trunk and synapse in the **superior**

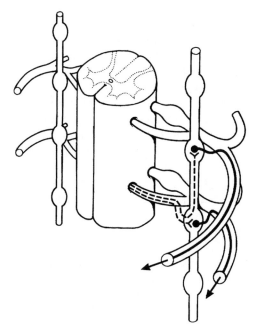

**Figure 1.15** Some postganglionic sympathetic nerves (solid lines) 'jump back' into mixed spinal nerves and run along with them to their destinations. Sympathetic nerves that supply sweat glands in the skin of the thorax travel in this way.

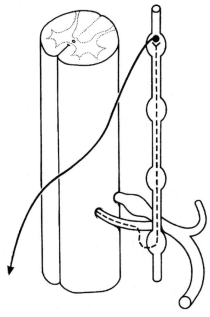

**Figure 1.17** Certain postganglionic sympathetic nerves (solid line) run free as nerves in their own right to their target organs. Sympathetic nerves to the heart, for example, run down through the neck as depicted in this figure.

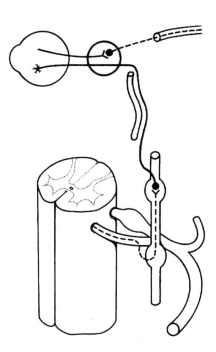

**Figure 1.16** Many postganglionic sympathetic nerves (solid line) run with blood vessels to their destinations. In this case they are also running through the parasympathetic ganglion (without synapsing of course) towards the eye.

**cervical ganglion.** Postganglionic fibres then leave the superior cervical ganglion and run along the internal or external carotid arteries and their branches towards target organs. These nerves follow the course shown in Figure 1.16.

Sympathetic nerve fibres destined for the upper limb originate at levels T3, T4, T5 and T6. They enter the sympathetic trunk and then run up it to synapse in the middle and inferior cervical ganglia, which are those nearest to the roots of the brachial plexus. From here the postganglionic sympathetic fibres (grey rami communicantes) run to join the mixed spinal nerves heading for the upper limb.

Sympathetic nerves destined for the heart, lungs or abdominal viscera arise between the levels of T1 and L2 as expected. Those to the heart and lungs run up to the middle and inferior sympathetic ganglia in the neck where they synapse. From here they 'run free' on their own as postganglionic sympathetic **cardiac** and **pulmonary nerves** in their own right (Fig. 1.17). These run down the neck again and mix together in the cardiac and pulmonary plexuses in the mediastinum. (A **plexus**, incidentally, is simply a mixing of nerves. The pleural of plexus is **plexuses**.)

Sympathetic nerves destined for the abdominal and pelvic viscera are a bit different. They are called sympathetic **splanchnic nerves**. They originate at segmental levels T5 to L2 and pass into the sympathetic trunk. Sympathetic splanchnic nerves are examples of nerves that break the generalization we spoke of and do not synapse close to the neuraxis in the sympathetic trunk. Instead these splanchnic nerves pass through the sympathetic ganglia *without* synapsing and 'run free' travelling towards their target organs, usually along blood vessels for some of the course. Sympathetic splanchnic nerves follow the course shown in Figure 1.14. Eventually, they synapse around the origins of major blood vessels arising from the aorta or, for example, in plexuses on the front of the sacrum or at the sides of the rectum in the pelvis in the so-called **hypogastric plexuses**. The blood vessels associated with sympathetic nerve plexuses include the coeliac trunk, superior mesenteric artery and inferior mesenteric artery. From these points of relay, the postganglionic sympathetic nerves run onwards, either with blood vessels or on their own to the abdominal or pelvic viscera. We will say more about these various plexuses later on.

# The parasympathetic nervous system

Parasympathetic nerves leave the neuraxis only from the cranial or sacral regions (Fig. 1.18). They are said to have a **craniosacral outflow** only. Parasympathetic nerve fibres leave the brain in the IIIrd, VIIth, IXth and Xth cranial nerves. They leave the sacral part of the spinal cord in segments S2, S3 and S4.

As in the sympathetic nervous system, the motor parasympathetic nerves are made up of two neurons that synapse in an autonomic ganglion. Parasympathetic ganglia, however, tend to be close to their target organs and a long way from the neuraxis. In the case of the heart, gut or bladder, for example, this synaptic relay actually takes place in the wall of the viscus itself. Other parasympathetic ganglia lie in plexuses close to their target organs (Fig. 1.19). In the floor of the pelvis, for example, there are plexuses where preganglionic parasympathetic nerves synapse and then relay postganglionic parasympathetic nerves onwards to the erectile tissue of the penis or clitoris. There is no parasympathetic nerve supply to either the upper or lower limbs. But the whole of the

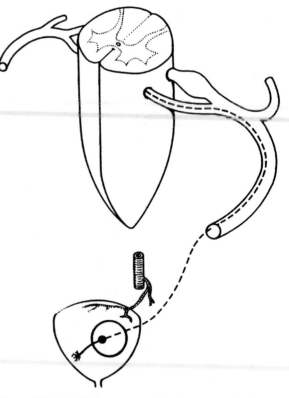

**Figure 1.19**  In this figure preganglionic parasympathetic nerve fibres (dotted line) are running out of the neuraxis through a sacral ventral nerve root and continuing on in the mixed sacral spinal nerve. The parasympathetic ganglion in this case is in the wall of the bladder and the postganglionic nerve fibre (solid line) is very short. Blood vessels are also running to the bladder.

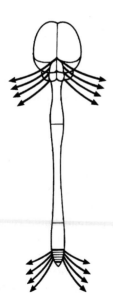

**Figure 1.18**  Parasympathetic motor nerves only ever leave the neuraxis in cranial nerves III, VII, IX and X or via the ventral nerve roots of sacral segments S2, S3 and S4.

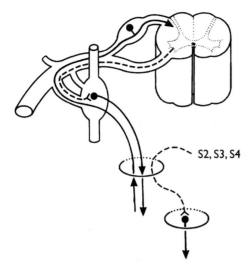

**Figure 1.20** Motor parasympathetic fibres arise in sacral segments S2, S3 and S4, and run to the parasympathetic ganglion where they synapse. Motor sympathetic fibres arise in the lateral horn of the spinal cord and synapse in sympathetic ganglia of the sympathetic trunk. Visceral afferent fibres are, in this case, following the path of sympathetic fibres back to the spinal cord. They have only one neuron and have their cell body in the dorsal root ganglion like any other sensory nerve.

head and neck and trunk get their parasympathetic nerve supply from the craniosacral outflow. The vagus nerve (Xth cranial nerve) has an extensive distribution of parasympathetic nerve fibres which reach as far distally as the last third of the transverse colon. The sacral outflow reaches up out of the pelvis as far as the transverse colon where the vagal innervation to the gut ends. It also supplies the pelvic viscera and genitalia.

Preganglionic parasympathetic fibres that arise from sacral segments S2, S3 and S4 are called **pelvic splanchnic nerves**. This is very confusing and they should not be mixed up with sympathetic splanchnic nerves. In fact a splanchnic nerve is simply a nerve running on its own to a viscus and can be either sympathetic or parasympathetic. It would be less confusing if we consistently spoke of *parasympathetic* pelvic

splanchnic nerves arising from segments S2, S3 and S4, and of *sympathetic* splanchnic nerves that arise from segments T5 to L2. The details of both parasympathetic secretor motor and sympathetic motor nerves in the head and neck, thorax, abdomen and pelvis will be described further in each appropriate section of these volumes.

## Visceral sensory nerves

Sensory nerves from the viscera run back to the central nervous system alongside autonomic nerves. These afferent visceral nerves, however, are just like somatic sensory nerves in that they have only one neuron peripherally. Some of these sensory nerves conduct special sensations like taste back to the brain in cranial nerves. Other general visceral sensations include feelings of fullness or emptiness in the stomach, bowel or bladder. These sensations travel back in the vagus nerve and may be associated with reflexes such as hunger or vomiting. Visceral pain from the thorax and abdomen is associated with smooth muscle spasm. The afferent nerve fibres, which carry visceral pain from the thorax and abdomen, travel back to the spinal cord side by side with the motor sympathetic nerves (Fig. 1.20). When they reach the mixed spinal nerve they travel back through the dorsal root like any other sensory nerve. They also have their cell bodies in the dorsal root ganglion. Visceral pain from the prostate, bladder, rectum and uterus, and from the perineum, travels back with the pelvic parasympathetic splanchnic nerves to segments S2, S3 and S4 of the spinal cord. In other words they are simply afferent nerves that run side by side with, or in company with, the motor parasympathetic nerves. They also have their cell bodies in the dorsal root ganglia of segments S2, S3 and S4. We will discuss visceral pain in the thorax, abdomen and pelvis in more detail at the end of the appropriate chapters in this book.

# The Thoracic Cage, Lungs and Pleura

The walls, floor and roof of the thorax enclose a space called the **thoracic cavity**. The walls are strong but also mobile and flexible, and are therefore well adapted to perform the vital movements of respiration. The bones of the thoracic walls consist of the vertebral column behind and of 12 pairs of ribs both posteriorly and at the sides and the sternum in front.

The **sternum** consists of three parts, the **manubrium**, the **body** of the sternum and the **xiphi-** sternum (Fig. 2.1). The manubrium and body of the sternum are joined at the **manubriosternal joint**. This is a **secondary cartilaginous joint**, that is, one where where fibrous tissue intervenes between the the cartilage covering the ends of the bones. The manubriosternal joint lies between the synovial joints of the second costal cartilages. The manubrium is set at a slight angle to the body of the sternum when viewed from the side. This angle is called the **manubriosternal**

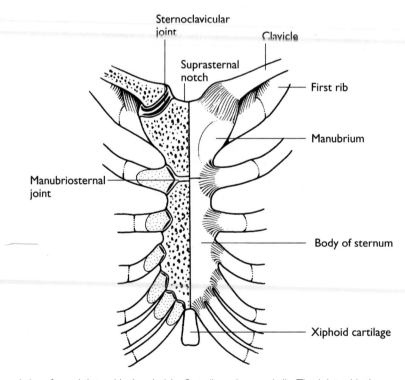

**Figure 2.1**  The manubrium forms joints with the clavicle, first rib and second rib. The joint with the second rib is at the level of the manubriosternal joint.

angle, or **angle of Louis** (a Parisian chest physician (1787–1872) who seems to have made his mark here). This prominence of the joint at the level of the second costal cartilage therefore forms a useful landmark, since the manubriosternal joint can always be felt in the midline. The second costal cartilage and rib can be confidently identified here. If you consider a subject standing in the anatomical position and viewed from one side (Fig. 2.2), then a horizontal line passing back from the manubriosternal joint to the vertebral column meets it at the level of the fourth thoracic vertebra (T4). This imaginary line defines the superior mediastinum above it. It is also the level of the bifurcation of the trachea during quiet respiration and of the concavity of the aortic arch. In addition it is the level at which the azygos vein joins the superior vena cava on the right. These are useful things to remember, since they help with the interpretation of computed tomograms of the chest, as we shall see later.

The vertebral column at the back of the thoracic cage consists of the bodies of the 12 thoracic vertebrae. (The anatomy of thoracic vertebrae is described in the section on the vertebral column in Volume 1 of this series.) Typically, each rib articulates posteriorly with the body of the vertebra of its own segmental level and with the one above as well. The synovial joint between the **head** of the rib and these two vertebrae therefore spans the intervertebral disc.

There are three exceptions to this: the first rib, 11th rib and 12th rib. These ribs articulate only with the vertebral body of their own segmental level. The **tubercle** of each rib always articulates with the transverse process of its own thoracic vertebra. This articulation is also a synovial joint, except for those between the tubercles of ribs 11 and 12, which are joined only by a ligament. Understand, therefore, that, by and large, movements of ribs during respiration involve synovial joints at both the head and tubercle of each rib and, in the case of the first seven ribs, with the sternum in front as well.

In front, the ribs becomes cartilaginous. This is because they ossify from posterior to anterior and fail to do so completely. The upper **costal cartilages** articulate with the sternum by means of synovial joints (Fig. 2.3). The exception is the first rib. This is firmly united to the manubrium with a **primary cartilaginous joint**, that is, a joint completely composed of cartilage. This means that the **thoracic inlet** (see below) is rigid in front and forms a firm base for muscles in the neck to act on during respiration and for the clavicle to lie on. The lower costal cartilages fuse together to form the **costal margin**. Ribs are sometimes classified as **true ribs, false ribs** and **floating ribs**. True ribs each meet the manubrium or sternum; false ribs join to form the costal margin; and floating

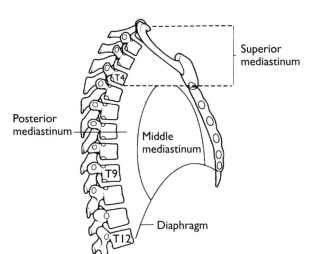

**Figure 2.2** The lungs lie on either side of the midline structures that make up the mediastinum. The heart forms the middle mediastinum. The structures above the level of T4 lie in the superior mediastinum. The structures behind the heart lie in the posterior mediastinum.

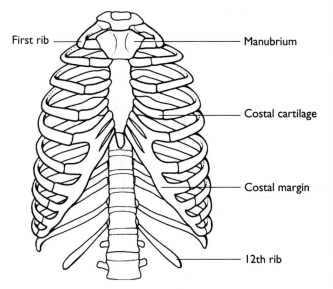

**Figure 2.3** The first rib joins to the manubrium at a primary cartilaginous joint. The remaining true ribs form synovial joints between their costal cartilages and the sternum. False ribs have costal cartilages that fuse together forming the costal margin. The last two floating ribs do not reach the costal margin.

ribs, the last two, are short and do not reach the costal margin at all.

The first rib lies flat but the other ribs are increasingly inclined so that, at the sides, they lie with their flat surfaces facing laterally and medially. On the undersurface of each typical rib there is a **costal groove**. **Intercostal nerves** and **vessels** run behind the costal groove (Fig. 2.4).

The thoracic cavity is closed off below by a domed muscle called the **diaphragm** (Fig. 2.2). In fact the diaphragm has two domes, one on the left and one on the right. Each of these rises high into the thorax on either side of the heart and mediastinum. The muscle of the diaphragm arises from the lower margins of the thoracic cage and the fibres sweep up towards a central tendinous portion. When the diaphragm contracts, the floor of the thoracic cavity therefore descends as the domes are drawn down together on the left and the right.

The roof of the thoracic cavity is closed over by a thin membrane called the **suprapleural membrane**. This sheet is in fact all that is left of a small muscle (scalenus minimus) but it remains, as the muscle was, anchored to the tubercle of the 7th (last) cervical vertebra. From here it spreads out to the margin of the first rib like a tent and is firm enough to resist balloon-ing upwards or being sucked downwards during respiration. The space bounded by the first thoracic vertebra, first rib and manubrium is referred to as the **thoracic inlet**. The **thoracic outlet** is closed over by the diaphragm below. We will say more about the diaphragm later and about movements of the ribs and sternum during respiration.

# Intercostal spaces

Intercostal muscles fill the spaces between the ribs (Fig. 2.4). They are active during respiration, especially inspiration of course, although the precise role of each group during inspiration and expiration is not clear. They certainly have an important role in preventing the intercostal spaces from 'imploding' during deep inspiration when there is negative atmospheric pressure in the thorax. They are also active during heavy lifting and straining, and resist bulging outwards when the intrathoracic pressure is raised. When they are paralysed, therefore, both of these do occur.

The **external intercostal muscles** extend from the sharp lower border of one rib to the more rounded upper border of the rib below (Fig. 2.5). The fibres of the muscles run downwards and forwards. Anteriorly, the muscle fibres become membranous; the **anterior intercostal membrane**. Deep to the external intercostal muscles are the **internal intercostal muscles**. The fibres of these muscles run downwards and backwards. Posteriorly, these muscle fibres fade away and are represented by the **posterior intercostal membrane**.

The innermost layer of muscle in the thoracic region is incomplete and less important functionally. As a group they are the **innermost intercostal muscles** or **transversus thoracis** muscles. Behind the sternum fibres of this group fan out rather than forming a complete sheet here. In other places, close to the vertebral bodies posteriorly and more laterally in the thoracic wall, muscle fibres of this group do not attach to each rib and cross more than one intercostal space. The **neurovascular plane** in which arteries, nerves and veins travel round the body wall is between the internal intercostal muscles and the innermost intercostals.

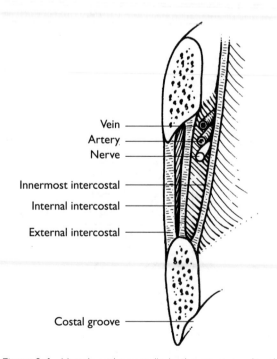

Vein
Artery
Nerve

Innermost intercostal

Internal intercostal

External intercostal

Costal groove

**Figure 2.4**  Vessels and nerves lie in the neurovascular plane between the internal and the innermost intercostal muscles.

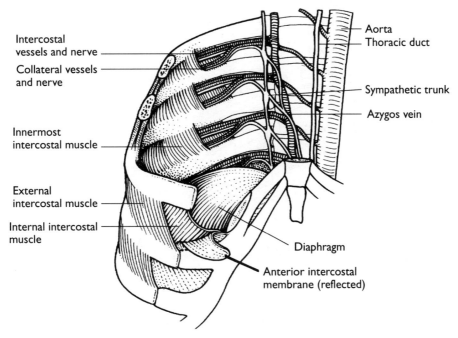

**Figure 2.5**   The direction of each group of intercostal muscle fibres and membranes alternate from external to innermost layers as they pass between ribs above and below. (After Last RJ. *Anatomy Regional and Applied.* Edinburgh: Churchill Livingstone, 1972.)

# Nerves, arteries and veins in the intercostal spaces

Mixed spinal nerves emerge from the intervertebral foramina between adjacent thoracic vertebrae. These divide into a ventral and dorsal ramus. In the thoracic region the ventral ramus, or intercostal nerve, gains the neurovascular plane of its corresponding intercostal space. Intercostal nerves give off collateral branches which supply sensation to the parietal pleura, the periosteum of the ribs, and which supply motor fibres to the muscles of the intercostal spaces. The collateral branch runs 'unprotected' close to the top of each rib in an intercostal space. The main intercostal nerve lies undercover of the costal groove and gives off a lateral cutaneous branch and a terminal anterior cutaneous branch. These branches pierce the body wall and supply the overlying skin. They also carry sympathetic fibres with them which innervate sweat glands, vessels and erector pili muscles in the skin.

It is worth knowing that pus arising from serious infections from the region of the vertebral column (such as were once common in cases of tuberculosis, for example) tends, in an intercostal space, to track along the path of the neurovascular structures. Abscesses erupt, or point, on the skin at the positions where the lateral and anterior cutaneous nerves and vessels pierce the body wall.

**Posterior intercostal arteries** arise segmentally from the aorta as it runs through the posterior mediastinum (Fig. 2.6). **Anterior intercostal arteries** arise from the **internal thoracic arteries** which run on either side of the sternum. Posterior and anterior intercostal arteries meet and anastomose. They both give off collateral branches which join and run with the collateral nerves close to the top of the rib below. Since the upper two intercostal spaces lie above the level of the descending aorta, they get their arterial supply separately from the **costocervical trunk** in the neck.

The **vena azygos** or **azygos vein** (Fig. 2.7) ascends into the thorax through the aortic opening behind the diaphragm in company with the aorta and the thoracic duct. The azygos vein receives most of the venous blood from the thoracic wall on the right side directly by means of **posterior intercostal veins**. The uppermost intercostal vein on the right drains into the brachiocephalic vein directly via the first **posterior intercostal vein**.

On the left, the majority of posterior intercostal

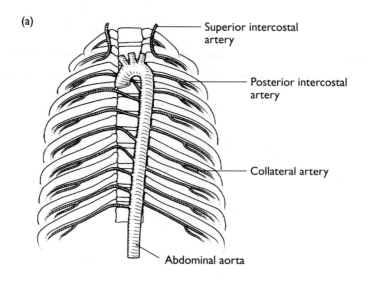

(a)

— Superior intercostal artery

— Posterior intercostal artery

— Collateral artery

Abdominal aorta

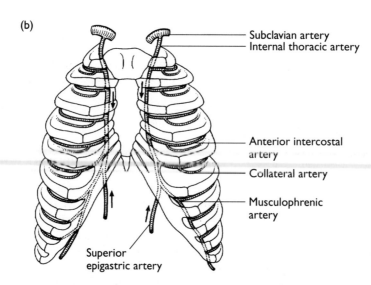

(b)

— Subclavian artery
— Internal thoracic artery

— Anterior intercostal artery

— Collateral artery

— Musculophrenic artery

Superior epigastric artery

**Figure 2.6**    Posterior and anterior intercostal arteries arise in a segmental manner from the aorta posteriorly (a) and the internal thoracic artery anteriorly (b). These vessels and their collateral branches anastomose with each other and with other arteries such as the superior epigastric artery in the rectus sheath.

veins drain into the **hemiazygos veins**. The first three posterior intercostal veins on the left, however, usually form a **superior intercostal vein** and drain directly into the left brachiocephalic vein in the superior mediastinum. Usually, the **superior hemiazygos vein** drains the 4th to 8th posterior intercostal spaces and the **inferior hemiazygos vein** drains the 9th to 12th spaces. Both hemiazygos veins then pass across the vertebral column to join the azygos vein on the right. At its upper end, the azygos vein arches forwards over the right main bronchus to join the superior vena

cava. We noted earlier that this junction is at the level of a line joining the manubriosternal joint and the lower border of the 4th thoracic vertebral body.

Now that we have described the vessels and nerves that supply the thoracic cage we can re-examine a cross-section through an intercostal space (Fig. 2.8). Notice the order of the vein, artery and nerve (VAN) under cover of the costal groove and also the position of the collateral vessels and nerves at the bottom of each space. The order of the collaterals is reversed from top to bottom, that is, NAV. Be sure you recall

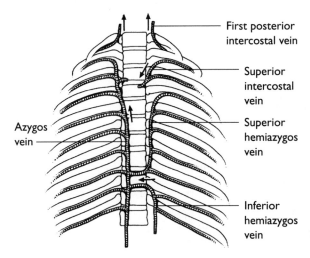

**Figure 2.7** Posterior intercostal veins drain into the azygos venous system. The uppermost spaces drain into the brachiocephalic veins direct or, in the case of the left second and third spaces, via the superior intercostal vein.

First posterior intercostal vein

Superior intercostal vein

Superior hemiazygos vein

Inferior hemiazygos vein

Azygos vein

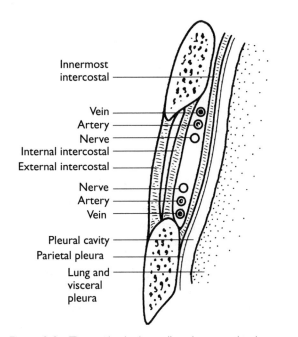

Innermost intercostal

Vein
Artery
Nerve
Internal intercostal
External intercostal

Nerve
Artery
Vein

Pleural cavity
Parietal pleura
Lung and visceral pleura

**Figure 2.8** The parietal pleura lies deep to the innermost intercostal muscles. The pleural cavity, visceral pleura and lung lie deep to this.

the neurovascular plane between the internal and the innermost intercostal muscles and understand that it is incomplete in places.

# The diaphragm

Three layers of muscle make up the body wall. The innermost layer originally lay inside the ribs. The middle layer of muscle passed between each of the ribs, and the outer layer lay outside the ribs. Phylogenetically, the diaphragm is part of the innermost layer of body wall muscle. Developmentally, the muscle that forms the diaphragm arises in the neck region. It drags its nerve supply with it during development, mostly from the ventral ramus of the 4th cervical spinal nerve. We will study its nerve and blood supply later on but it makes sense to complete our study of the thoracic cage by looking at the muscular anatomy of the floor of the thorax now.

We saw earlier that the adult diaphragm is a double-domed structure forming the floor of the thoracic cavity and the roof of the abdominal cavity. The muscle fibres originate as slips from the inner rim of the costal margin and posterior abdominal wall (Fig. 2.9). They arch upwards to insert into a tendinous central portion of the diaphragm called the **central tendon**. In the midline anteriorly, fibres arch upwards to the central tendon from the xiphisternum. More laterally, fibres arise from the insides of the 7th to 12th costal cartilages. The posterior origin of muscle fibres is a little more complicated. Here the diaphragm arises from two fibrous arches on either side. These are best seen on the posterior abdominal wall. The arches are called the **medial** and **lateral arcuate ligaments**. They are thickenings of fascia that arch over muscles of the posterior abdominal wall. Their presence allows these muscles to move freely and independently behind the diaphragm. The details of these ligaments and their relationships will be studied together with the musculature of the posterior abdominal wall but, basically, the muscular fibres arising from the medial and lateral arcuate ligaments run upwards towards the central tendon of the diaphragm. In the midline posteriorly, there are other muscles that contribute to the diaphragm. These arise from the vertebral column in the form of two muscular bundles. They are called the **right** and **left crus**. The right crus arises from the 1st, 2nd and 3rd lumbar vertebral bodies and the discs between them. The left crus arises from only the 1st and 2nd lumbar vertebral bodies and the disc between these.

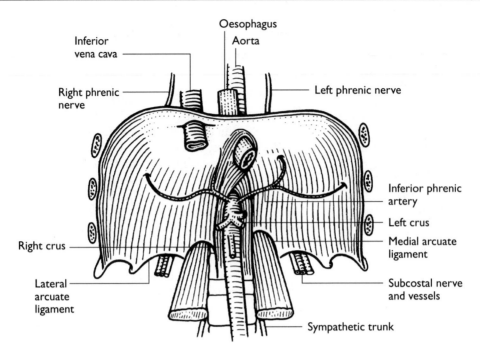

Inferior
vena cava

Right phrenic
nerve

Oesophagus

Aorta

Left phrenic nerve

Inferior phrenic
artery

Left crus

Medial arcuate
ligament

Right crus

Lateral
arcuate
ligament

Subcostal nerve
and vessels

Sympathetic trunk

**Figure 2.9** The diaphragm separates the thoracic and abdominal cavities. The aorta, inferior vena cava, oesophagus and phrenic nerves each pass through the diaphragm.

## Structures that pass between the thorax and abdomen

Various structures pass either behind the diaphragm or in front of it, or pierce it. The descending aorta enters the abdomen though the **aortic opening** at the level of T12. It is guarded on either side by the two crura (plural of crus). The aorta really lies behind the diaphragm, which ensures that the vessel is not constricted when the muscular fibres of the diaphragm contract. In fact, there are three structures that use the so-called aortic opening for transit between abdomen and thorax. These are the aorta, the thoracic duct through which lymph flows from the abdomen to the thorax, and the vena azygos.

Some arterial blood reaches the anterior abdominal wall via the internal thoracic arteries. These two arteries descend on the inner surface of the costal cartilages on either side of the sternum. At the xiphisternum, they divide into musculophrenic and superior epigastric arteries. The **superior epigastric** arteries pass *between* the anterior muscular fibres of the diaphragm, on either side, to reach the anterior abdominal wall. The **musculophrenic** arteries supply the muscle of the diaphragm anteriorly, as well as the

pericardium. Posteriorly, the **inferior phrenic** arteries arise directly from the aorta and supply the posterior aspects of the diaphragm from beneath.

The inferior vena cava pierces the central tendon of the diaphragm to the right of the midline at the level of T8. This opening in the diaphragm is called the **caval opening**. It is surrounded by tendon, which ensures that the vein is not compressed when the diaphragm contracts. In fact, muscle pull on the central tendon draws it taut and therefore helps to keep the inferior vena cava patent during inspiration.

The oesophagus pierces the muscular part of the diaphragm to the left of the midline at the level of the 10th thoracic vertebral body. It passes through muscular fibres of the left crus here, but fibres of the right crus also cross over the midline and encircle the oesophageal opening like a sling. As the diaphragm contracts this helps to prevent regurgitation of food back into the oesophagus. It is a back-up mechanism to the slower-acting smooth muscle of the cardiac sphincter.

## Nerves that pierce the diaphragm

Several nerves pierce the diaphragm. The phrenic nerves are mixed spinal nerves but provide the only motor nerve fibres to supply the muscle of the diaphragm. They originate in the neck from segmental levels C3, C4 and C5 (mostly C4). The course of the phrenic nerves through the thorax will be described below but, on reaching the upper surface of the diaphragm, the left phrenic nerve pierces the diaphragm on its own. On the right, the phrenic nerve accompanies the inferior vena cava through the diaphragm. The phrenic nerves then spread and ramify on the undersurface of the diaphragm and innervate it. The phrenic nerves are also sensory nerves to the parietal peritoneum, parietal pleura and pericardium overlying the diaphragm. The lower intercostal nerves (T7 to T12) also give sensory proprioceptive branches to the periphery of the diaphragm.

The two vagus nerves enter the abdomen in front of and behind the oesophagus. The sympathetic trunks pass behind the medial arcuate ligaments. As we have seen, the sympathetic trunks in the thorax give off three preganglionic nerves, the **greater, lesser** and **least splanchnic nerves**. These reach the abdomen by piercing the crura on either side.

## Movements of the ribs and diaphragm during respiration

Deep inspiration results in an increase in all diameters of the thorax. The main muscle of inspiration is the diaphragm. As the diaphragm contracts it is drawn down and the volume of the thoracic cavity enlarges. During quiet inspiration there is an increase in the anteroposterior diameter of the thorax as the sternum moves forwards. This movement has been likened to that of a pump handle (Fig. 2.10). There is also an increase in the transverse diameter of the thorax as the ribs rise. The movement of true ribs during inspiration (which you will remember attach directly to the sternum) is described as being analogous to the movement of a bucket handle. The false ribs, that attach to the costal margin, spread apart in a manner analogous to a pair of spreading calipers during inspiration. Expiration follows as there is relaxation of all the respiratory muscles and recoil of thoracic cage. There is also elastic recoil of the lungs as the thoracic cavity reduces in volume. During deep and forced inspiration many other accessory muscles of respiration act on the chest cage. The scalene muscles in the neck pull up on the first and second ribs. When

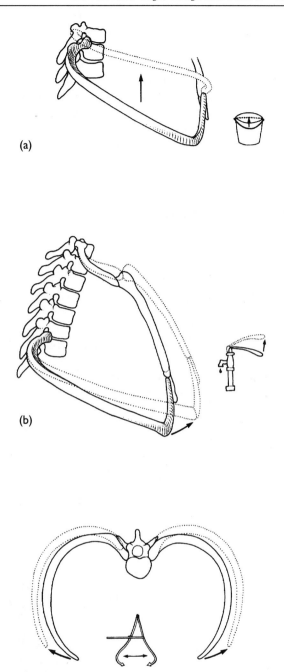

(a)

(b)

(c)

**Figure 2.10** Synovial joints between the ribs and sternum allow movements in different directions during respiration. At the sides they swing out and up in the way that a bucket handle moves (a). In front the manubrium moves out and up like the handle of a stand pump (b) and lower down the ribs also spread like measuring calipers (c). (After Aiello LC and Dean MC. *Human Evolutionary Anatomy.* London: Academic Press, 1990.)

the upper limbs are fixed, the pectoral muscles can act on the rib cage to elevate the ribs. Even small accessory muscles of respiration surrounding the nasal aperture act to improve respiratory efficiency. It is also possible to expire forcefully. In this situation the muscles of the abdominal wall are active and raised intra-abdominal pressure forces the abdominal viscera against the diaphragm and so up and into the thoracic cavity.

## The lungs and pleural cavity

The lungs are elastic and are constantly expanding and recoiling during inspiration and expiration. They are surrounded by a double-layered sac called the **pleura** which enables them to do this within the thoracic cavity. The developing lungs grow into the pleura as if they were pressing into a soft balloon (Fig. 2.11). The outer layer of **parietal pleura** lies against the inner walls of the thoracic cavity. The inner layer of **visceral pleura** covers the surface of the lungs, even dipping into the fissures that divide the lung tissue.

The **pleural cavity** lies within the two layers of pleura and contains only a thin film of fluid which allows free movement of the lungs.

Structures enter and leave the lung at the **hilum** or **root of the lung**. We will study the root of the lung in more detail later when the topography of the lungs, heart and great vessels has been described. The two layers of pleura reflect back on themselves at the hilum and become continuous. They form a kind of cuff or sleeve here, which is slack at rest and sags below the hilum to form the **pulmonary ligament**. During exercise, increased venous return from the lungs swells the pulmonary veins. Since these lie in the lower portion of the lung root, they expand and take up the slack in the pulmonary ligament.

Parietal pleura lines the inside of the chest wall, the lateral aspects of the mediastinum and the superior aspect of the diaphragm. **Endothoracic fascia** binds the parietal pleura firmly to the inner aspect of the rib cage and to the diaphragm but not to the mediastinum. Superiorly, parietal pleura runs over the superior aperture of the thorax as the **dome of the pleura** to line the **suprapleural membrane**. Notice that

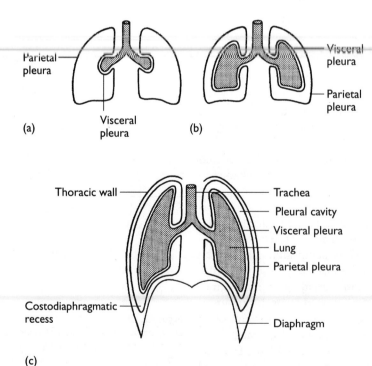

*Figure 2.11* The developing lungs grow into the pleural cavity and become surrounded by visceral and parietal pleura. These two layers of pleura enclose a potential space called the pleural cavity. (After Snell RS. *Clinical Anatomy for Medical Students.* Boston: Little, Brown, 1981.)

parietal pleura extends right down into the deep, narrow gutter between the chest wall and diaphragm. This is called the **costodiaphragmatic recess**. Even in deep inspiration the lung itself does not occupy the depths of the costodiaphragmatic recess. The parietal pleura extends down as far as T12 posteriorly, T10 in the mid-axillary line and T8 anteriorly in the mid-clavicular line (Fig. 2.12). The lungs themselves extend only to about two intercostal spaces short of this, i.e. to T10 posteriorly, to T8 in the mid-axillary line and to T6 anteriorly.

## Nerve supply to the pleura

The parietal pleura is sensitive and is supplied by the collateral branches of the intercostal nerves in a segmental manner around the chest wall. The diaphragmatic and mediastinal parietal pleura are supplied by the phrenic nerves. Visceral pleura is insensitive and supplied only by vasomotor autonomic nerves.

## The structure of the lungs

The lungs are roughly conical in shape, and adapt to fit the shape of the pleural cavities (Fig. 2.13). The **diaphragmatic surface** is concave since it rests on the dome of the diaphragm. Around the edge of the diaphragm the lung, covered by its visceral pleura, pushes only a little way down into the costodiaphragmatic recess. It therefore has a crescentic lower edge. The **mediastinal surface** is that part of the lung which moulds itself against the mediastinum. It is into this surface that structures enter and leave the lung though the hilum.

The right and left principal bronchi can easily be identified in the roots of the right and left lungs, by the cartilages in their walls. Notice, however, that on the right side, the principal bronchus and the right pulmonary artery start to branch before entering the lung tissue. The pulmonary veins are thin walled and able to expand so that they can cope with variation in blood flow. Remember that during exercise, for example, they carry a lot of blood and are therefore dilated. They lie in the lower part of each hilum in the pulmonary ligament.

The mediastinal surface of the lungs relate through pleura to the structures of the mediastinum. On the

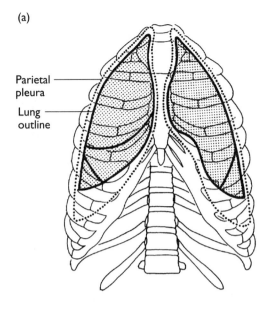

(a)

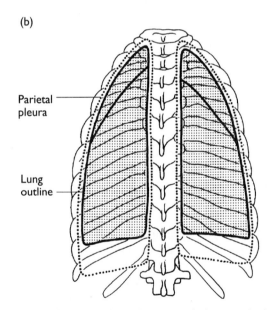

(b)

**Figure 2.12** The parietal pleura extends down to the level of the 12th rib posteriorly, the 10th rib laterally and the 8th rib anteriorly. The lungs and visceral pleura lie two rib spaces above this level. (a) Anterior and (b) posterior view of the rib cage, lungs and pleura.

right side the main structures related to the lungs are 'venous'. These structures are the superior vena cava, the right atrium and the inferior vena cava. On the left side the lung is related to 'arterial' structures, namely the great arteries, the arch of and the descending aorta and left ventricle. In fact, the

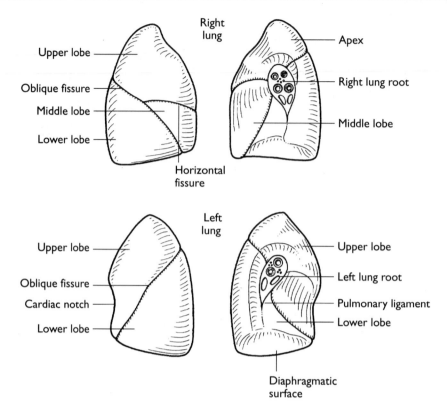

**Figure 2.13** Oblique fissures divide both the right and left lungs into upper and lower lobes. An additional horizontal fissure in the right lung creates a middle lobe.

bulging left ventricle prevents the lung from reaching the midline. The left lung therefore has a concavity on its mediastinal surface called the **cardiac notch**. The apex of the lung and the pleura rise above the level of the superior aperture of the thorax and lie against the suprapleural membrane.

The lungs are divided completely by **fissures** which extend over the surface to the hilum (Fig. 2.13). These fissures divide the lungs into **lobes**. On both sides a long **oblique fissure** runs around the lung. The surface marking of this fissure starts at the level of the spine of the 3rd thoracic vertebra posteriorly and ends in front at the level of the 6th costal cartilage in the mid-clavicular line. As a rough guide, with an arm placed on the head, the surface marking of this fissure then corresponds posteriorly to the slope of the medial border of the scapula. The oblique fissure cuts right into the lung tissue and so completely divides each lung into an **upper lobe** and a **lower lobe**.

In the right lung there is an additional fissure called the **horizontal fissure**. This extends laterally from the level of the 4th right costal cartilage anteriorly to the oblique fissure. It also cuts in deeply as far as the hilum, and so demarcates another lobe, the **right middle lobe**. The right lung therefore has three lobes and the left lung only two. The fissures and lobes can be used, therefore, to identify the lungs.

During growth, the right lung has to push up and out from beneath the azygos arch. If it does not clear the arch completely, the vein cuts off another lobe. This lobe is called the **azygos lobe**.

# Bronchopulmonary segments

The right and left principal bronchi both divide into **lobar bronchi**. Each lobar bronchus supplies a lung lobe (Fig. 2.14). Therefore, there are three lobar bronchi on the right and two on the left. On the right side, the lobar bronchus to the upper lobe arises outside the lung tissue. Inside the lung, the middle and lower lobar bronchi branch off. On the left side, the division is simply into upper and lower lobar bronchi. Each

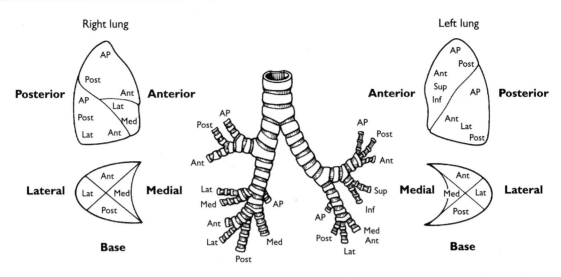

**Figure 2.14** There are three lobar bronchi in the right lung and two in the left. Each of these branches into a segmental bronchus that supplies a named bronchopulmonary segment. In the right lung the apical, posterior and anterior bronchopulmonary segments lie in the upper lobe. The lateral and medial segments lie in the middle lobe and the apical, basal lateral, medial, posterior and anterior segments lie in the lower lobe. On the left the pattern is the same except that the upper lobe contains superior and inferior bronchopulmonary segments, equivalent to the middle lobe. Note that the apical segmental bronchus of the lower lobe is oriented posteriorly. In a supine patient there is therefore poor drainage from this bronchopulmonary segment.

lobar bronchus divides to supply definite segments of the lobe. These branches are called **segmental bronchi** and each supplies a named **bronchopulmonary segment**. Their importance is that, often, disease in the lung can be localized to one or two of these segments. Different bronchopulmonary segments are also dependent, or susceptible, to fluid accumulating in them when a person lies on their back, side or sit upright. This has important implications for patient care.

The pattern of the bronchopulmonary segments is not too difficult. In the upper lobe, the lobar bronchus divides into three segmental bronchi: an **apical**, an **anterior** and a **posterior**. These supply air to three similarly named bronchopulmonary segments. On the left side the apical and posterior segments are combined and so only two segmental bronchi are needed. These are then called the **apicoposterior** and **anterior segmental bronchi**. The left upper lobe bronchus also supplies a 'tongue' of lung tissue in the lower part of the upper lobe. It corresponds to the middle lobe of the right lung, but is not divided off clearly by a fissure. This part of the upper lobe is called the **lingula** and is composed of two bronchopulmonary segments. They are supplied by **superior** and **inferior lingular segmental bronchi**. On the right side, the middle lobe is clearly demarcated by the horizontal fissure. It therefore has its own lobar bronchus. This divides into

**lateral** and **medial segmental bronchi** which supply similarly named bronchopulmonary segments.

The lower lobes on both sides are the same. Each is supplied by a **lower lobe lobar bronchus**. This divides into five segmental bronchi: an **apical** segmental bronchus, and **anterior**, **posterior**, **lateral** and **medial basal** segmental bronchi. Occasionally, the medial basal segment on the left is greatly reduced or absent (because the heart encroaches on the lung here) and it shares a segmental bronchus with the anterior basal segment.

# Blood supply, nerve supply and lymphatic drainage of the lung

The lung tissue receives blood supply from the **bronchial arteries**, branches of the aorta, and blood drains from the lungs through **bronchial veins**. These vessels belong to the systemic circulation. Unlike the bronchial arteries the bronchial veins do not closely follow the segmental bronchi but rather also run in the intersegmental septa.

Autonomic nerves mix in the pulmonary plexuses, which are found mostly on the posterior aspects of the principal bronchi. Sensation from both the mucous membrane and stretch receptors in the lung tissue, as well as parasympathetic secretor motor nerves run in the vagus nerve. These spread along the bronchial tree. There are also vasomotor sympathetic nerves to the bronchial arteries and sympathetic efferents that produce bronchodilatation.

Lymphatic drainage from the lung tissue and bronchi is plentiful. The drainage is towards the hilum where there are **hilar nodes**. From these, drainage goes on towards the **mediastinal nodes**. The lymphatic drainage of the lungs is often the pathway by which carcinoma cells disperse from their primary site in the lung tissue or bronchus and into the mediastinum.

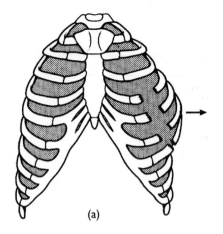

(a)

# Applied anatomy of the thoracic cage and lungs

Injuries to the chest wall can result in fractured ribs. These are extremely painful during normal respiratory movements but are especially so when coughing or sneezing. More serious injuries to the chest wall that involve fractures to several ribs can result in a large portion of the chest wall moving freely and independently from the rest. In this situation the 'free' section moves inwards on inspiration and outwards on expiration (Fig. 2.15). There is said to be **paradoxical movement** of this portion of the rib cage. The condition is known as **flail chest**. Clearly, when this seriously impairs ventilation it affects oxygenation of the blood. In this situation, free-moving segments must be fixed so that they cannot compromise respiration in this way.

The sternum is an important site of erythropoiesis and is also, being subcutaneous in the midline, easily accessible for **sternal puncture**. In this procedure bone marrow is drawn off for examination with a wide-bore needle from the medullary cavity of the sternum under local anaesthesia.

When it is necessary either to draw off fluid from the pleural cavity or to tap air from a pneumothorax, a needle is inserted between two ribs into the pleural cavity. There is a danger, when doing this in the region of the costodiaphragmatic recess, that the diaphragm will be pierced and the abdomen entered in error. By choosing, say, the 6th or 7th intercostal space, in the mid-axillary line, or in the mid-scapular line, this complication can be avoided. Because the intercostal nerve and vessels are not always com-

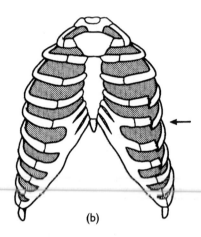

(b)

**Figure 2.15**  Injuries to the chest wall may separate sections that then move freely and paradoxically during respiration. On inspiration the section is sucked in; on expiration the section bulges out.

pletely protected behind the costal groove, and since the collateral branches run close to the top of the rib beneath, there is a risk of neurovascular trauma at these sites. This can be avoided, or at least minimized, by trying to avoid the lower or upper borders of ribs in an intercostal space.

The visceral pleura is insensitive to pain but the parietal pleura is very sensitive. Irritation or injury to the costal or peripheral diaphragmatic parts of the pleura results in pain that is well localized over the site, although it may well radiate further anteriorly along the same dermatome. (A fractured 10th rib may give pain as far anteriorly as the umbilicus.) The mediastinal and central diaphragmatic parietal pleura are supplied by the phrenic nerve. Pain from these regions is often referred to the root of the neck and shoulder. This happens because the cutaneous innervation of the neck and shoulder derives from the same spinal segments that give rise to the phrenic nerve. Disease in the lung tissue, however, may not irritate the parietal pleura and so may not be painful. Alternatively, it may present late in this way.

Understanding the arrangement of bronchopulmonary segments enables different portions of the lung to be drained by placing patients on one side or the other, or on their front or back (or even upside down for a short while). It follows that ill or unconscious patients benefit from being moved frequently since this promotes good aeration and drainage of the lungs.

# The Heart

A midline partition called the **mediastinum** divides the thoracic cavity into right and left spaces. These spaces are occupied by the right and left lungs. The mediastinum itself can also be divided into sections. The bulk of the mediastinum is made up of the heart and, since this more or less occupies the middle section of the mediastinum, the region outlined by the heart is called the **middle mediastinum**.

In front, between the heart and the sternum, there is a small narrow space called the **anterior mediastinum**. Above the heart, and lying behind the manubrium, is the **superior mediastinum**. It contains the roots of the great vessels of the heart. At the back, between the heart and the vertebral column, is the **posterior mediastinum**. To begin with, we will describe the anatomy of the heart and then return to study the anterior, superior and posterior mediastinum in the next chapter.

## The heart

The heart is a four-chambered muscular pump that squeezes oxygenated blood into the aorta and deoxygenated blood into the pulmonary trunk. The heart is contained in a sac called the **pericardium**. You will remember that the four chambers are the right and left **atria** and the right and left **ventricles**. Notice before we proceed any further that the cavities do not really lie 'right' and 'left' in the thorax (Fig. 3.1). The lies heart obliquely across the thorax and, in addition to this, the right side is turned to face the front. Seen from the front, therefore, the surface of the heart consists mainly of the anterior walls of the right atrium and right ventricle (Fig. 3.2). A small part of the left

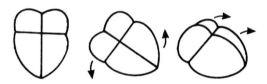

**Figure 3.1**  The heart lies obliquely in the mediastinum and is rotated over to show very little of the left side when seen from the front.

atrium and the left ventricle forms the left edge and apex of the heart in this view.

The ventricles present a different pattern from the atria. The atria are collecting chambers from the systemic and pulmonary circulations. They are thin walled. The ventricles on the other hand are thick-walled muscular pumps. The pressure is greater on the left side of the heart in the adult, since it takes much more effort to pump blood all the way around the systemic circulation than through the pulmonary circulation. The left ventricle is therefore larger and thicker.

To begin with, let us look in general at the four chambers of the heart and their inflow and outflow channels. Inflow to the heart is into the right and left atria. Blood drains into the right atrium by means of the **superior vena cava** from the upper parts of the body and through the **inferior vena cava** from below (Fig. 3.3). Inflow into the left atrium comes from the lungs via the **pulmonary veins** (Fig. 3.4). The left atrium and pulmonary veins lie, for the most part, at the back of the heart. They are therefore best seen by looking at the posterior surface of the heart. This surface is made up almost completely by the left atrium. The walls of the left atrium overlap the left ventricle

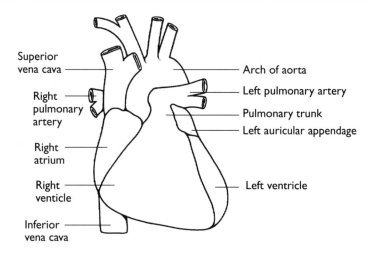

Superior vena cava — Arch of aorta

Right pulmonary artery — Left pulmonary artery

— Pulmonary trunk

— Left auricular appendage

Right atrium —

Right venticle — Left ventricle

Inferior vena cava —

**Figure 3.2** The heart and great vessels seen from the front.

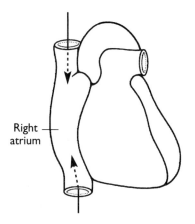

Right atrium —

**Figure 3.3** Blood returns to the right atrium via the superior and inferior vena cava.

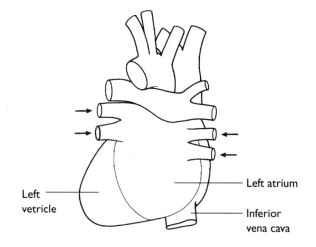

Left atrium —

Left vetricle —

— Inferior vena cava

**Figure 3.4** Blood returns from the lungs to the left atrium via the four pulmonary veins (arrowed). The left atrium is seen from behind here.

on one side and a narrow strip of the right atrium on the other. The heart rests on the central tendon of the diaphragm and this diaphragmatic surface is composed entirely of right and left ventricular walls.

Next, notice the direction of blood flow from the right atrium into the right ventricle (Fig. 3.5). Now look at the outflow channels of both the right and left ventricles (Fig. 3.6). These are the **pulmonary trunk** and the **aorta** respectively. These two structures spiral around and cross each other. Notice the direction of blood flow from the ventricles into these vessels. Notice now that the inflow–outflow angle is more acute in the left ventricle than in the right ventricle.

This sort of general appreciation of the topography of the heart in the thorax is exactly what is needed when interpreting the outline of the heart and great vessels on a normal posteroanterior chest radiograph (Fig. 3.7). The left border is formed from above downwards by the aortic arch, the left pulmonary artery, the left auricular appendage and the left ventricle. The aortic arch shows as a rounded projection which is called the **aortic knuckle** on a radiograph. The right border of the heart on a radiograph is formed by a hint of the inferior vena cava, visible when the radiograph is taken in deep inspiration, the superior vena cava and the right atrium.

Having established some basic facts about the heart we now need to look at each chamber of the heart in more detail. We will follow them through from the

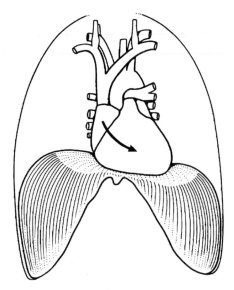

**Figure 3.5** The heart lies over the central tendinous portion of the diaphragm and the great vessels lie behind the manubrium. The arrow indicates the direction of blood flow from the right atrium through to the right ventricle.

right atrium in the same order that blood circulates through the heart.

## The right atrium

This chamber is best seen from the front. It is capped by the right **auricular appendage**. If the front wall of

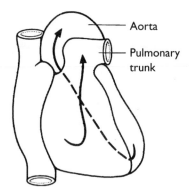

**Figure 3.6** The arrows represent the direction of blood flow out of the left and right ventricles and into the aorta and pulmonary trunk respectively.

the atrium is opened, part of the interior is seen to be roughened by raised horizontal ridges of muscle called **musculi pectinati** (Fig. 3.8). The rest of the interior of the right atrium is smooth walled. The junction between the rough and smooth parts is called the **crista terminalis**. On the outside of the heart there is sometimes, but not always, a groove that corresponds to this ridge called the **sulcus terminalis**. These ridges and the two types of lining of the atrium reflect their different developmental histories.

In the embryonic heart, two chambers, the **sinus venosus** and the **common atrium**, are separated by a groove. The sinus venosus enlarges and forms the greater part of the right atrium. It is represented in the adult by the smooth-walled part of the adult right atrium. The small remaining piece of common atrium

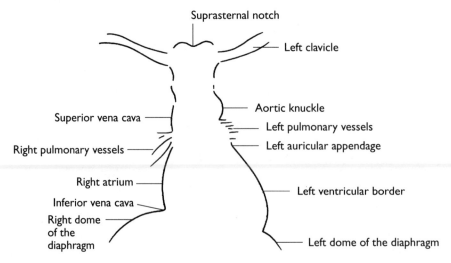

**Figure 3.7** Outline of the heart and great vessels as seen on a posteroanterior chest radiograph. Start with the aortic knuckle on the left of the patient and work your way round the labelled structures. Refer back to Figures 3.2 and 3.5 as you do this.

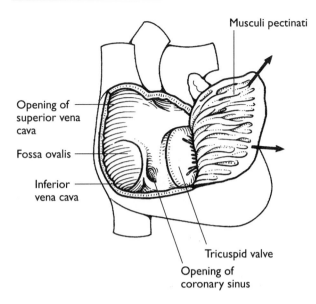

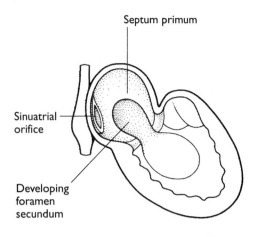

**Figure 3.8** The superior and inferior vena cava and the coronary sinus open into the right atrium. The smooth wall between the atria bears the fossa ovalis. The anterior part of the wall is raised by horizontal and parallel musculi pectinati.

**Figure 3.9** Within the primitive atrium there is downgrowth of the septum primum and development of a foramen secundum. (After Fitzgerald MJT and Fitzgerald M. *Human Embryology*. London: Baillière Tindall, 1994.)

on the right is represented by the rough-walled part of the adult right atrium. The part of the primitive common atrium into which the pulmonary veins drain is gradually separated off from the right atrium by an interatrial septum and eventually becomes the primitive left atrium. Almost all of the left atrium is smooth internally because it is derived from the primitive pulmonary veins. Only the left auricle derives from the original common atrium and so it remains the only part of the adult left atrium that is ridged with bands of muscle internally.

Look next at the wall that separates the right atrium from the left atrium (Fig. 3.8). It bears scars that reflect the developmental history of this septum. Initially, a primary septum (the **septum primum**) grows down from the roof of the common atrium (Fig. 3.9) but then breaks down superiorly to form a foramen (the **foramen secundum**). The lower part continues to grow down and finally divides the common atrium into two chambers. Then a second septum (the **septum secundum**) grows down from the roof and covers over the foramen secundum (Fig. 3.10). However, it too fails to close off the two atria completely and leaves a crescentic **foramen primum** low down opposite the septum primum. There is then a hole high up in the septum primum and a hole low down in the septum secundum. Blood can

easily flow obliquely through these holes between the left and right atria because the septa remain unfused before birth.

The fetus has no need of lungs or a pulmonary circulation. It receives oxygenated blood from the placenta via the umbilical vein. Oxygenated blood from the placenta enters the fetus through the umbilical cord and eventually reaches the inferior vena cava and the right atrium. Since there is little point in sending oxygenated blood through non-functioning lungs, a fold of endocardium, which forms the rudimentary valve around the opening of the inferior vena cava, directs blood towards the partly formed interatrial septum. On reaching the atrial septum blood is driven obliquely through the oval foramen and foramen secundum (which act as a flap-like valve) into the left atrium. At birth the lungs begin to function and the pressure in the left atrium rises. The septum primum and septum secundum are held shut and eventually fuse.

The scars of this process can still be seen in adult atrial septum (Fig. 3.8). In the adult, the depressed **fossa ovalis** represents the septum primum of the developing heart. Its surrounding rim, the **limbus fossa ovalis** or **annulus**, represents the lower free edge of the septum secundum.

Venous blood drains into the right atrium from three areas: the upper part of the body, the lower part of the body and the heart muscle itself. The **superior vena cava** drains blood from the upper part of the body, the **inferior vena cava** drains blood from the

(a)

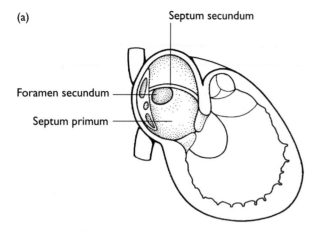

Septum secundum

Foramen secundum

Septum primum

(b)

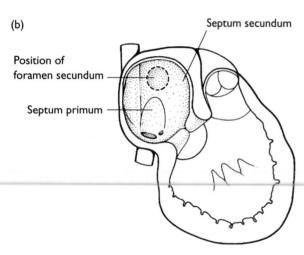

Septum secundum

Position of foramen secundum

Septum primum

**Figure 3.10**   The scars on the wall of the adult right atrium result from the septum primum becoming covered by the septum secundum, except for the region within the fossa ovalis. (After Fitzgerald MJT and Fitzgerald M. *Human Embryology.* London: Bailliere Tindall, 1994.)

lower part of the body, and a vein called the **coronary sinus** drains the majority of blood from the heart muscle itself. The inferior vena cava and the coronary sinus have rudimentary valves at their openings into the right atrium but the superior vena cava has none. The venae cavae lie in line with the right border of the heart and open into the right atrium above and below respectively. The opening of the coronary sinus lies near the septal cusp of the tricuspid valve.

Blood leaves the right atrium through the right atrioventricular orifice. This is guarded by the **right atrioventricular valve** or **tricuspid valve**.

# The right ventricle

This chamber lies mostly towards the front of the heart. The walls are very thick and muscular but, because the left side is so much thicker, the septal wall between the two ventricles bulges into the right ventricular cavity (Fig. 3.11). This gives a crescentic shape to the cross-section of the right ventricle. When opened along its length, the cavity of the right ventricle is triangular in shape. The inflow is down into the posteroinferior angle of the triangle and the outflow is up through the posterosuperior angle. The inner walls of the ventricle are lined with muscular ridges called the **trabeculae carneae** (Fig. 3.12). One of these trabeculae lies free, and crosses the cavity of the ventricle as a muscular rod from **septal** to **anterior wall**. It is called the **septomarginal trabecula**, or sometimes the **moderator band**. It is important because it contains a bundle of specialised conducting tissue through which impulses spread to the wall of the right ventricle.

The right atrioventricular orifice is surrounded by a fibrous ring which provides attachment for the **right atrioventricular** or **tricuspid valve**. This is a one-way valve. To prevent reflux or regurgitation of blood back into the right atrium, the valve has three **cusps** which point into the ventricle at all times. The cusps are named **anterior**, **posterior** and **septal**, by virtue of their positions. Their bases are attached to the fibrous ring, and their edges are tethered to the ventricular wall by a series of delicate fibrous strands called **chordae tendineae**. Tension in the chordae tendineae is transferred to the cusp edges. The chordae tendineae

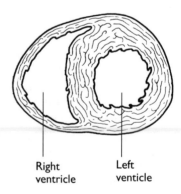

Right ventricle        Left ventricle

**Figure 3.11**   A section through the walls and cavities of the right and left ventricles reveals that right ventricle is crescentic in cross-section and that the wall of the left ventricle is three times thicker than that of the right.

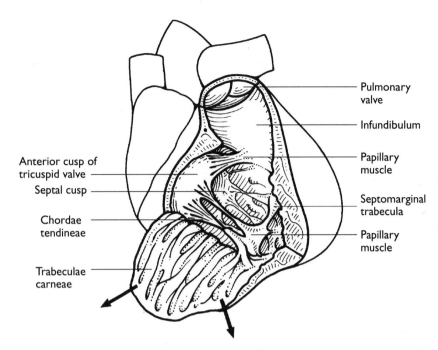

**Figure 3.12** The tricuspid valve guards the right atrioventricular orifice. The septomarginal trabecula and trabeculae carneae create rough ridges on the right ventricular wall. The infundibulum beneath the pulmonary valve is smooth walled.

attach to three conical muscular projections of the ventricular wall which anchor the chordae. These are the **anterior**, **posterior** and **septal papillary muscles**, but the chordae arising from each do not all insert into the correspondingly named cusp. Instead, the chordae from each papillary muscle insert into the edges of adjacent cusps. The pull of the chordae and the contractions of the papillary muscles therefore draw their adjacent edges together and into the ventricle. This arrangement prevents the cusps being driven back into the atrium and turned inside out during ventricular contraction.

The outflow from the right ventricle leads into the pulmonary trunk. This rises from the posterosuperior angle of the ventricle. As the opening of the trunk is approached, the wall of the ventricle becomes smooth and conical. This part of the ventricle is called the **infundibulum**. (This simply means a funnel.) The pulmonary trunk is guarded by a different kind of valve composed of three cusps or **semilunar valvules**. This is the **pulmonary valve** (Fig. 3.13). Its **valvules** are markedly convex towards the ventricle and concave towards the trunk. At the centre of each valvule is a **fibrous nodule**. The three valvules are named from

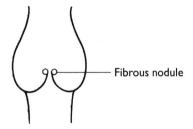

**Figure 3.13** Cross-section through two of the three semilunar valvules that guard the pulmonary trunk.

the positions they occupy: **anterior**, **right** and **left**. Regurgitation of blood back into the ventricle is prevented simply by the shape of the valvules. There are no chordae or papillary muscles in the pulmonary trunk. Ballooning of the cusps under back pressure simply presses their opposing surfaces together and shuts off any back flow. The wall of the pulmonary trunk opposite each valvule is slightly dilated to form a **sinus**.

The part of the ventricular wall that separates the right ventricle from the left is called the **interventricular septum**. Most of the interventricular septum is muscular, but understand that in its posterior aspect, just between the atrioventricular orifices and where it is in line with the interatrial septum, it is thin and membranous. Originally, in the embryo, there is a common ventricle with a common outflow channel, but various septa grow into the cavity such that it becomes completely divided into right and left ventricles and into aortic and pulmonary outflow channels. We will mention some of the important anomalies that occur in this region later.

## The left atrium

Blood returns from the lungs to the left atrium via the pulmonary veins. Only the auricular appendage of the left atrium can be seen on the front of the heart. Because of the oblique and rotated position of the heart in the body, the left atrium is best seen from behind. As we have seen, only the auricular appendage of the left atrium is trabeculated; the rest of the cavity is smooth (Fig. 3.14). As the pulmonary vein is 'taken in' to form the wall of the atrium during development, its branches get nearer and nearer to the atrial wall. Eventually, the first four branches end up entering the posterior wall of the atrium separately. This is the usual adult pattern. The four pulmonary veins bring oxygenated blood from the lungs into the left side of the heart. Blood then leaves the left

atrium through the left **atrioventricular orifice**. This is guarded by the left **atrioventricular valve** or **mitral valve**.

## The left ventricle

The left ventricle is thicker than the right since it has to work against the higher resistance of the systemic circulation. Its internal surface is raised by trabeculae carneae, in the same way as on the right side. However, they are more numerous in this ventricle (Fig. 3.15). Blood arrives in the ventricle through the left atrioventricular orifice. This is smaller than the right atrioventricular orifice but, like it, is surrounded by a fibrous ring, just big enough to admit two fingertips. The left **atrioventricular valve**, or **mitral valve**, is also attached to the fibrous ring. The mitral valve is built on the same plan as the tricuspid valve except that it has only two cusps, an anterior and a posterior cusp. As in the case of the tricuspid valve, chordae tendineae are attached to the cusp edges. These sprout from just two papillary muscles in the left ventricle, an anterior and a posterior muscle. Each muscle sends chordae tendineae to each cusp, attaching to adjacent edges so that on ventricular contraction the cusps are drawn into the ventricle and together.

The outflow from the ventricle is through the **aortic**

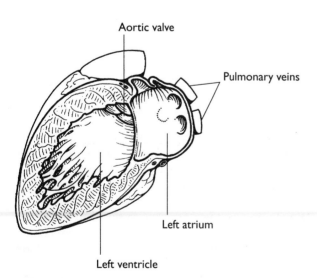

**Figure 3.14**   The left atrium is completely smooth walled except for the inner aspect of the auricular appendage. The four pulmonary veins empty into the left atrium. The two on the right are shown here.

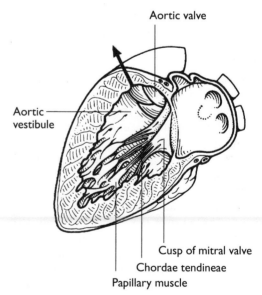

**Figure 3.15**   The mitral valve with its chordae tendineae and papillary muscles guards the left atrioventricular orifice. The wall of the aortic vestibule is smooth.

valve into the aorta. The stream of blood passes over the anterior wall of the anterior cusp of the mitral valve and into the smooth-walled **aortic vestibule**. From here it proceeds through the aortic valve. This valve consists of three valvules that are constructed in exactly the same way as the pulmonary valvules (Fig. 3.16). Their positions, however, are different, the valvules being **posterior**, **right** and **left** in position. The aorta swells into three sinuses immediately above the level of the valve, one opposite each valvule. The aortic valve works like the pulmonary valve. Following ejection of blood from the ventricle the major blood vessels recoil and pressure is then higher in the aorta than in the ventricle. The valvules balloon back into the aortic vestibule thus pressing the cusp surfaces together.

## The conducting system of the heart

Notice now that the fibrous rings of the four valves of the heart are continuous with each other (Fig. 3.17). They not only form the basis for the attachment of their corresponding valve cusps but also form an electrical barrier between the atrial and ventricular muscle of the heart. This fibrous meshwork surrounding the valve openings is called the **fibrous skeleton** of the heart.

Electrical impulses can spread through the heart muscle but not through the fibrous skeleton of the heart. Each beat of the heart is initiated in the right atrium at the upper end of the crista. The area where this occurs is called the **sinuatrial node** (SA node). From the SA node, rhythmic impulses pass through the atrial musculature, causing them to contract and discharge blood into the ventricles. It is likely that impulses spread through the right atrial wall in several specialized bundles, both through to the left atrium and towards the fibrous skeleton of the heart (Fig. 3.18). However, this is not universally accepted. The impulses do not pass directly through the skeleton of the heart and so the ventricles are relaxed while atrial contraction occurs. Impulses eventually reach the interatrial septal region near the opening of the coronary sinus. Here, just above this opening, close to the septal leaflet of the tricuspid valve, is another specialized group of myocytes called the **atrioventricular node** (AV node).

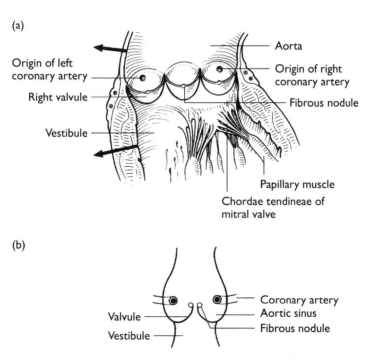

**Figure 3.16**   The aorta and left ventricle have been opened to show the aortic valve. The valvules each have cusps and a fibrous nodule which abut each other to prevent back flow of blood into the ventricle. The coronary arteries arise in the right and left aortic sinuses.

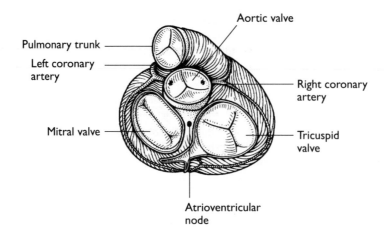

Aortic valve

Pulmonary trunk

Left coronary
artery

Right coronary
artery

Mitral valve

Tricuspid
valve

Atrioventricular
node

**Figure 3.17** The valves of the heart lie in the same plane in a fibrous skeleton, which prevents electrical impulses passing between the atria and ventricles.

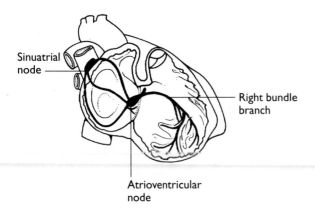

Sinuatrial
node

Right bundle
branch

Atrioventricular
node

**Figure 3.18** Electrical impulses are initiated and spread from the sinuatrial node through the atrial wall and towards the atrioventricular node.

Impulses from the AV node travel onwards through the **atrioventricular bundle** (of **His**). This short bundle pierces the fibrous skeleton and arrives in the region of the thinner membranous part of the interventricular septum (Fig. 3.19). In the interventricular septum, the bundle divides into **right** and **left crura**, or **bundle branches**, which pass to the respective ventricles. The left crus supplies the papillary muscles in the left ventricle and then spreads out as a network in the ventricular wall. The right crus takes impulses to the septal and posterior papillary muscles of the right ventricle. It then proceeds in the septomarginal trabecula to the anterior papillary muscle. Eventually it terminates and sends out many branches to form the **Purkinje network**.

Impulses reaching the AV node from the atria are delayed a little as they pass through the trunk and crura. The impulses first reach the papillary muscles and their contraction closes both atrioventricular valves. Further rapid spreading of the impulses causes simultaneous contraction of both ventricles, and blood is ejected into the pulmonary trunk and aorta. Although the heart can beat by itself, each stroke initiated at the SA node, this inherent activity is influenced by the autonomic nervous system. Sympathetic fibres to the heart carry impulses that produce an increase in the rate of impulses generated at the SA node. The heart beats faster. Parasympathetic activity on the other hand slows down the rate.

There is in fact a three-tier system of rhythmic activity in the heart. First, the muscle fibres themselves have a built-in capacity for contracting rhythmically. If the SA node is destroyed the muscle still contracts, although at a slower rate. The second tier of control is the SA node, which under normal circumstances dictates the frequency of contraction. Finally the autonomic nervous system influences the rate according to the particular requirements of the body for blood. One must not of course forget the direct effects of noradrenaline from the adrenal gland on the heart.

## The cardiac plexuses

Both sympathetic and parasympathetic nerve fibres contribute to the cardiac plexuses. The **superficial cardiac plexus** lies beneath the arch of the aorta. The **deep cardiac plexus** lies in front of the bifurcation of

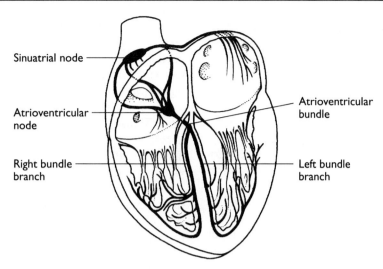

**Figure 3.19** The right and left bundle branches arise from the short atrioventricular bundle and spread through the walls of the ventricles.

the trachea. So in reality this is just one great mass of nerve fibres close to the heart. The sympathetic fibres to the heart arise in the upper four thoracic segments of the spinal cord (Fig. 3.20). From here they pass to the segmental ganglia in the sympathetic trunk and either synapse at this level or ascend to one of the cervical sympathetic ganglia and synapse

there. Postganglionic sympathetic fibres, therefore, converge on the cardiac plexuses both directly, from the upper four thoracic ganglia of the sympathetic trunk, and from the neck via cardiac branches from the cervical sympathetic ganglia. These nerve fibres pass through the cardiac plexuses (obviously, without synapsing here) and then travel on to the SA node.

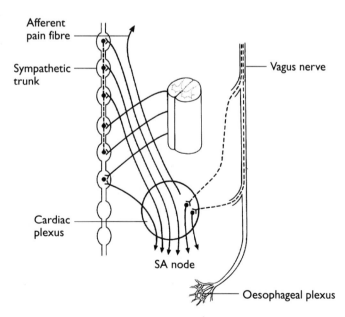

**Figure 3.20** Three of the four segmental preganglionic spinal sympathetic nerves that supply the heart are shown passing to the sympathetic trunk. These synapse either at this level or higher in the cervical sympathetic trunk, and postganglionic sympathetic fibres then pass into the cardiac plexus. Preganglionic parasympathetic fibres from the vagus nerve also pass to the plexus and synapse within it. From here all autonomic nerves innervate the sinuatrial (SA) node. Afferent pain fibres follow sympathetic fibres back to the dorsal roots of the spinal cord.

The parasympathetic input to the cardiac plexuses comes from the vagus nerve (the Xth cranial nerve). Cardiac branches arise in the neck and then run down to the plexuses (Fig. 3.20). They also arise from the recurrent laryngeal branch of the vagus nerve (see below) and pass directly from here to the cardiac plexuses. Preganglionic parasympathetic fibres either synapse in the cardiac plexuses or pass straight through them and synapse in the walls of the atria. There is then a stream of autonomic fibres passing from the cardiac plexuses to the SA node, the atrial walls and the coronary vessels. Afferent pain fibres from the heart follow both parasympathetic and sympathetic fibres back to the spinal cord. We will consider pain from the heart later in the applied anatomy section at the end of this chapter.

## The coronary arteries

The arterial blood supply to the heart comes from the root of the aorta by means of two **coronary arteries**. We saw earlier that the aortic valve has a posterior, a right and a left valvule. Just beyond this valve there are three corresponding aortic sinuses (posterior, right and left sinuses). The coronary arteries arise from the right and left aortic sinuses and are called the **right** and **left coronary arteries**. They both follow

a course around the heart, in the groove between the atria and ventricles, and meet each other again at the back of the heart. The groove is called the **coronary sulcus**. While in the sulcus they give off branches to both atria and ventricles.

The right coronary artery arises from the right aortic sinus and squeezes between the right auricular appendage and the infundibulum to reach the coronary sulcus (Fig. 3.21). As it proceeds in the sulcus it gives branches to the right atrium and right ventricle. These include branches to the SA node and one to the pulmonary trunk. The largest of these branches runs along the right margin of the heart and is called the **marginal branch**. It supplies the right ventricular wall. The right coronary artery now turns on to the posterior surface of the heart and gives off its largest branch. This is called the **posterior interventricular artery**. It travels along the **interventricular groove** on the diaphragmatic surface of the heart towards the apex of the heart. The right coronary artery supplies all of the right atrium and most of the right ventricle. It usually also supplies the AV node and AV bundle. The part *not* supplied by it is a strip adjoining the interventricular groove on the front of the heart. The posterior interventricular branch, however, *does* supply a corresponding strip to the left ventricle of the diaphragmatic surface. This branch also supplies the posterior part of the interventricular septum. The

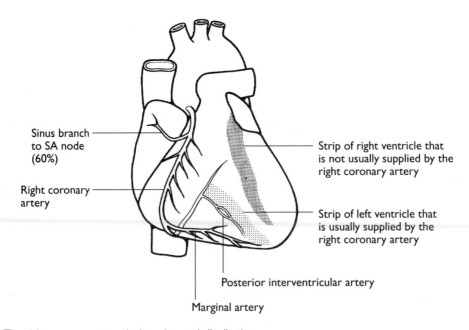

Sinus branch to SA node (60%)

Right coronary artery

Strip of right ventricle that is not usually supplied by the right coronary artery

Strip of left ventricle that is usually supplied by the right coronary artery

Posterior interventricular artery

Marginal artery

**Figure 3.21**    The right coronary artery, its branches and distribution.

arterial branch to the SA node arises from the right coronary artery in almost 60% of cases. In 90% of cases the artery to the AV node arises from the posterior interventricular branch of the right coronary artery. The AV bundle usually has dual blood supply from both arteries.

The left coronary artery arises from the left aortic sinus. It squeezes between the left auricular appendage and the infundibulum to reach the coronary sulcus (Fig. 3.22). Notice that this is only a short distance and that the artery proper is therefore not very long. Once in the sulcus it divides into two terminal branches. A large **anterior interventricular artery** travels towards the apex in the anterior interventricular groove. At the apex its terminal branches anastomose with those of the posterior interventricular artery. The other terminal branch is the **circumflex artery**. This continues around the heart in the coronary sulcus. It anastomoses with terminal branches of the right coronary artery. The left coronary artery supplies the left atrium and left ventricle (except for the small strip of left ventricle on the diaphragmatic surface). In addition, the anterior interventricular branch also supplies the strip of right ventricle adjacent to it and the anterior part of the interventricular septum. In the septum it anastomoses with branches of the posterior interventricular artery.

The coronary arteries are the only effective arterial supply to the heart muscle. They travel along the surface of the heart in grooves. As they proceed they give

branches into the myocardium (Fig. 3.23). There is little anastomosis between the main arteries and only a little between their branches. Blockage of one of the arteries is serious and although there are anastomoses, for instance between the two interventricular branches, these are often not large enough to form a collateral circulation quickly enough following sudden occlusion of one coronary artery.

## The cardiac veins

Venous blood collects in veins, which mainly drain along the grooves on the surface of the heart. To a lesser extent, veins drain directly into the cavities of the heart on the right side. The pattern of the surface veins is variable, but eventually they all drain towards

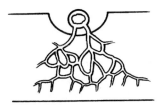

**Figure 3.23** The coronary arteries lie in grooves on the surface of the heart. Their branches run into the myocardium beneath.

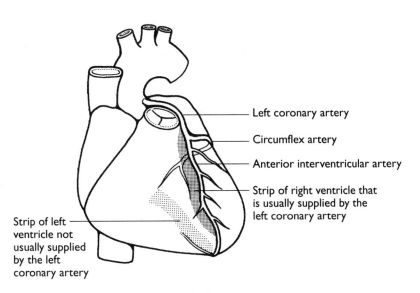

Left coronary artery

Circumflex artery

Anterior interventricular artery

Strip of right ventricle that is usually supplied by the left coronary artery

Strip of left ventricle not usually supplied by the left coronary artery

**Figure 3.22** The left coronary artery, its branches and distribution.

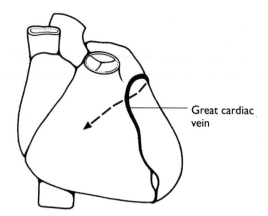

**Figure 3.24**   The great cardiac vein runs in the anterior inter-ventricular groove and then in the left part of the coronary sulcus where it becomes the coronary sinus.

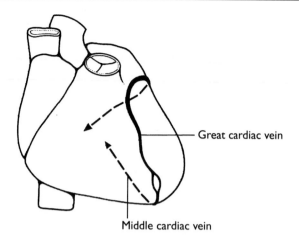

**Figure 3.25**   The middle cardiac vein approaches the coronary sulcus from the posterior interventricular groove.

the coronary sulcus on the posterior surface of the heart.

The **great cardiac vein** follows the anterior inter-ventricular groove and then the left part of the coron-ary sulcus (Fig. 3.24). It receives venous blood from atrial and ventricular tributaries. The **middle cardiac vein** travels along the posterior interventricular groove (Fig. 3.25). It also arrives at the coronary sul-cus on the back of the heart. The **small cardiac vein** runs with the marginal artery as far as the coronary sulcus and then also proceeds to the back of the heart (Fig. 3.26). A few so-called **anterior cardiac veins** drain directly into the right atrium. All three large veins (great, middle and small) drain into the **coron-ary sinus**. This is found on the posterior surface of the heart in the coronary sulcus. It has already been seen

draining into the right atrium. Some blood from the heart muscle drains directly into the chambers of the heart by means of small myocardial veins called **venae cordis minimae**. Excess tissue fluid is drained via lymphatics which follow the coronary arteries to lymph nodes found in the mediastinum.

# The pericardium

The heart is contained in a sac called the **pericardium**. The outer layer is called the **fibrous pericardium**. It is thick and tough and blends with the outer layer of the roots of the great vessels above. Below it is fused to the central tendon of the diaphragm. In front the

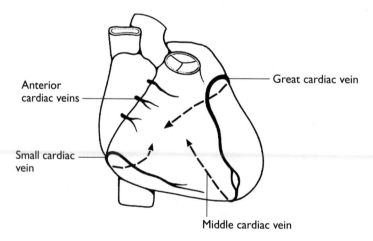

**Figure 3.26**   The small cardiac veins run with the marginal artery and back to join the coronary sinus in the coronary sulcus. The anterior cardiac veins drain directly into the right atrium.

fibrous pericardium is connected by fibrous tissue to the upper and lower parts of the sternum. These are the so-called **sternopericardial ligaments**.

Within the fibrous pericardium there is a serous pericardial sac. It is as if the heart has been pressed into a balloon contained within the fibrous pericardial sac. The double-layered serous pericardial sac envelops the heart and comes to be applied both to the inner wall of the fibrous pericardium and to the outer aspect of the heart. The inner layer of serous pericardium is in contact with the heart muscle and is called the **visceral pericardium**. The outer layer is firmly adherent to the fibrous pericardium and is called the **parietal pericardium**. Between the two layers is a pericardial cavity that contains a thin film of fluid. This allows the heart free frictionless movement within the pericardium.

At the roots of the great vessels the serous pericardium runs up them for a short way like a sleeve and then reflects back on itself. The superior vena cava has a sleeve of its own but the aorta and the pulmonary trunk share a sleeve. A finger within the pericardial cavity can therefore encircle the aorta and pulmonary trunk together and emerge between them and the superior vena cava. This is called the **transverse sinus** of the pericardial cavity (Fig. 3.27). Two fingers introduced into the pericardial cavity behind the left atrium of the heart are reflected back around

the inside of the inner serous pericardial sac. This posterior recess in the pericardial cavity is called the **oblique sinus** of the pericardial cavity.

# Applied anatomy of the heart and pericardium

The surface markings of the outline of the heart are shown in Figure 3.28. The position of the apex of the heart corresponds to the position of the **apex beat** of the heart in adults and can usually be felt in the left 5th intercostal space close to the mid-clavicular line. Heart sounds provide an invaluable guide to the condition of the heart valves. Because the valves are positioned so close together, the sounds from each valve are not clearly distinguishable from each other directly over their surface markings. **Auscultatory** sites for each valve are positioned widely apart and allow the sounds produced by one valve to be clearly distinguished from those of the other valves. The sound from a heart valve is carried in the direction of blood flowing though the valve. Look at the auscultatory sites illustrated in Figure 3.28 and try to relate these to the surface markings of the valves themselves and to the direction of blood flow through them.

The heart is very sensitive to ischaemia (but not at all to touch, cutting or temperature). The nerve impulses from the heart that are responsible for relaying pain from the myocardium run back to the upper four or five thoracic segments on the left side. They do this in company with motor sympathetic nerve fibres travelling in the opposite direction. Pain from the heart is **referred** from these segments to the farthest site along the dermatome from the spinal cord. This is a general rule that explains many examples of referred pain in the body. In the case of the heart this is over the central region of the sternum, where it is felt as a tight crushing pain. Heart pain can also be felt across the left side of the chest and in the left arm. The reason for this is that segments T1 and T2 also contribute to the medial cutaneous nerve of the arm. There are also, occasionally, commissural neurons within the spinal cord that transmit impulses to equivalent segments on the right side of the spinal cord. It follows that referred pain from the heart may present on both sides of the chest together and rarely even in the right arm as well.

Pain from the pericardium may also be felt beneath

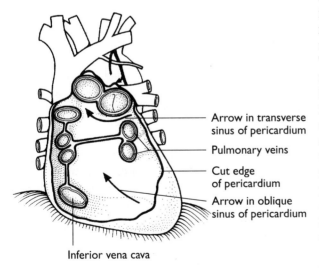

Arrow in transverse
sinus of pericardium

Pulmonary veins

Cut edge
of pericardium

Arrow in oblique
sinus of pericardium

Inferior vena cava

**Figure 3.27** A hand placed behind the heart in the pericardial cavity is reflected back on itself around the oblique pericardial sinus. A finger placed behind the aorta and pulmonary trunk on the left emerges on the right through the transverse pericardial sinus.

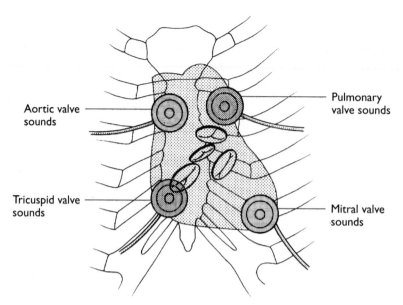

Aortic valve sounds

Pulmonary valve sounds

Tricuspid valve sounds

Mitral valve sounds

**Figure 3.28**   Sounds from each of the heart valves are maximal and most isolated from each other at the auscultatory points indicated.

the sternum and can result from fluid collecting in the pericardial cavity (**pericardial effusion**). Inflamed pericardial surfaces rub against each other and sound like a 'rustle' on auscultation with a stethoscope. Greatly increased pressure due to fluid accumulation in the pericardial sac can embarrass heart action. Fluid must then be drawn off the pericardial cavity through the 5th or 6th intercostal space on the left, close to the sternum. In this position it is possible to do this without piercing the pleural cavity. Take another look at Figure 2.12 to confirm this. This procedure is called **paracentesis** of the pericardium.

The central tendinous portion of the diaphragm is fused with the fibrous pericardium and develops from the same embryological source. The diaphragm develops from three main sources: the **septum transversum**, which represents the fused cervical myotomes of the 3rd, 4th and 5th cervical segments; the **pleuroperitoneal folds**, which are raised from the body wall as the pleura develop; and the **dorsal mesentery** of the oesophagus which contributes to the central portion of the diaphragm. Congenital hernia of the diaphragm results most commonly when there is failure of fusion between the septum transversum and the left pleuroperitoneal fold. In these cases infants may be born with abdominal organs herniated

through into the thoracic cavity which then compresses the lung and displaces the heart.

There are many congenital anomalies of the heart but it is worth mentioning a few here that build on the account of the anatomy of the heart given above. ASD, or atrial septal defect, is common. Some 25% of normal hearts have a patent foramen ovale where a small probe can be passed through from the right atrium to the left. This of course represents the incomplete adhesion of the septum primum and septum secundum, and is usually of no clinical significance. More serious defects in the fossa ovalis result in a considerable amount of deoxygenated blood from the right atrium entering the left atrium. In these situations the ASD requires repair. **VSD**, or ventricular septal defect occurs most often in the membranous part of the interventricular septum. The complex embryological origins of the interventricular septum may explain why this is the most common of all heart anomalies. Several congenital anomolies of the aorta and pulmonary trunk occur. These include transposition of the two great arteries, the development of a common arterial trunk, or development of a greatly diminished pulmonary trunk. Combinations of all these anomalies often occur. Perhaps the most well known is the **tetralogy of Fallot**.

chapter
4

# The Anterior, Superior and Posterior Mediastinum

## The anterior mediastinum

The anterior mediastinum is a thin, narrow space lying in front of the heart. In the adult it contains only a little fatty tissue and the remnants of the thymus gland. This gland is concerned with the production of lymphocytes during childhood when it is large and covers the great veins in the superior mediastinum and the percardium of the middle mediastinum.

## The superior and posterior mediastinum

Since several structures span both the superior and the posterior mediastinum, it makes sense to study these together. The superior mediastinum lies above the heart, above a horizontal line joining the manubriosternal joint to the lower border of the body of the 4th thoracic vertebra (T4). It helps to think about the structures in the superior mediastinum as being arranged in layers, from the most superficial structures anteriorly to the deepest posteriorly. When the manubrium is removed the superior mediastinum first comes into view. Superficially, the first structures to be seen are veins. These veins arise in the head and neck and in the upper limbs, and all drain to the superior vena cava (Fig. 4.1).

The **internal jugular veins** come from the head. These meet the **subclavian veins** from the upper limbs in the superior mediastinum. The internal jugular and

subclavian veins on either side join to form **right** and **left brachiocephalic veins**. These are asymmetrical as they run behind the manubrium. The right brachiocephalic vein passes directly downwards behind the right border of the manubrium to the superior vena cava. The left vein has to pass obliquely across the back of the manubrium to reach the superior vena cava.

Directly beneath this superficial layer of veins lie the great arteries that supply the head, neck and upper limbs (Fig. 4.2). These great vessels arise from the arch of the aorta. The arteries are also asymmetrical at their origins. The first ascending branch of the aorta is the **brachiocephalic artery**. This divides into the **right subclavian artery** and the **right common carotid artery**. The **left common carotid artery** and the **left subclavian artery** arise separately from the arch of the aorta. The brachiocephalic and left common carotid arteries clasp the **trachea**.

The trachea runs in the next layer, deep to the great vessels (Fig. 4.3). The trachea divides into **right** and **left principal bronchi** behind the bifurcation of the pulmonary trunk, at the level of the manubriosternal joint (T4). Finally, lying directly behind the trachea is the oesophagus. The aorta and the oesophagus run through both the superior mediastinum and the posterior mediastinum. The aorta descends through the posterior mediastinum on the left and then crosses to the midline **aortic opening** behind the diaphragm. The oesophagus runs behind the heart in the midline but then veers to the left as it approaches the oesophageal opening in the diaphragm. The two structures therefore cross in the posterior mediastinum (Fig. 4.4). On to this broad plan we need to add

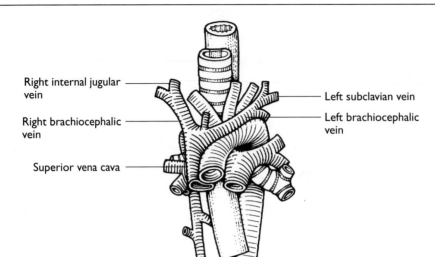

**Figure 4.1**   The right and left brachiocephalic veins are formed as the subclavian and internal jugular veins meet. The brachiocephalic veins join to form the superior vena cava on the right. These great veins lie immediately behind the manubrium in the superior mediastinum.

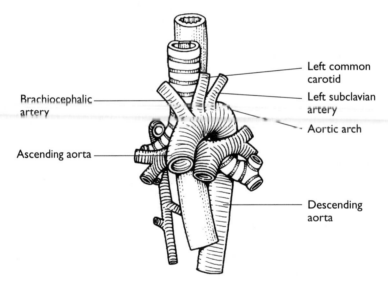

**Figure 4.2**   The ascending aorta and the arch of the aorta with its three major branches lie beneath the great veins. The brachiocephalic artery and the left common carotid artery clasp the sides of the trachea.

details about each important structure in the mediastinum.

# The phrenic nerves in the thorax

The right and left **phrenic nerves** arise in the neck from the ventral rami of the 3rd, 4th and 5th cervical spinal nerves (mostly the 4th). On the right side of the thorax notice that the phrenic nerve runs along the great venous channels, against the superior vena cava, then over the pericardium overlying the right atrium and finally on to the inferior vena cava (Fig. 4.5). In this way it reaches the diaphragm. On the left side the nerve follows the lateral surface of arterial structures. First it runs over the subclavian artery, then against the left common carotid artery and finally on to the arch of the aorta and pericardium overlying the left ventricle.

(a)

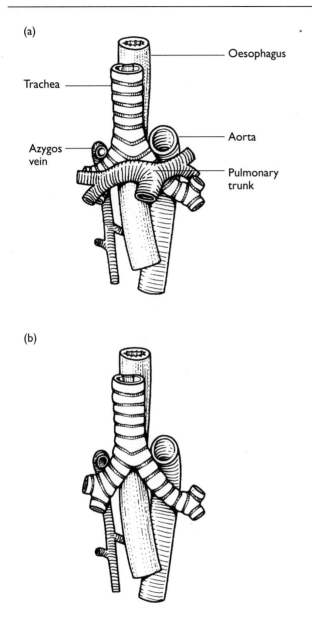

(b)

**Figure 4.3** The pulmonary trunk divides beneath the arch of the aorta into the right and left pulmonary arteries. The trachea lies between the oesophagus posteriorly and the arch of the aorta anteriorly in the superior mediastinum.

Both phrenic nerves pass in front of the roots of the lungs. They pierce the diaphragm at the level of T8 and innervate it from beneath. The phrenic nerve on the right passes through the diaphragm close to the inferior vena cava. The left phrenic nerve passes through on its own. Remember that the phrenic nerve is a mixed nerve. It is the only motor nerve to

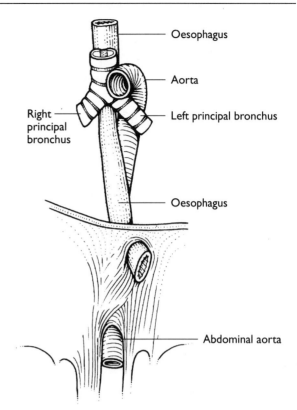

**Figure 4.4** The descending aorta and the oesophagus cross each other in the posterior mediastinum before they pass through the diaphragm.

the muscle of the diaphragm. But it is also sensory to the pericardium, parietal pleura and peritoneum overlying the diaphragm. Other sensory nerve fibres invade the diaphragm peripherally from the lower intercostal nerves (T7 to T12) which cross the costal margin.

## The trachea

The trachea is kept patent by incomplete cartilaginous rings that hold the lumen open. Posteriorly, the ends of each tracheal cartilage are joined by a few involuntary muscle fibres called the **trachealis muscle** (supplied by parasympathetic fibres from the vagus nerve). In this way the size of the lumen is under autonomic control. The trachea runs in the midline of the neck from the lower border of the larynx to the thoracic inlet. It is important to check that this is always so, both by direct examination and radiographically, since any deviation to one side may be

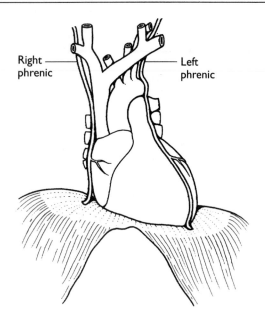

**Figure 4.5**   On the right the phrenic nerve lies against the superior vena cava, then the right atrium before piercing the diaphragm. On the left the phrenic nerve lies against the subclavian and common carotid arteries before crossing the arch of the aorta and running within the pericardium over the left ventricle.

indicative of a space-occupying lesion on the contralateral side. The trachea lies behind the manubrium in the superior mediastinum but then travels deep to the great veins and arteries. It ends at the level of the manubriosternal joint by dividing into **right** and **left principal bronchi**. The trachea is an elastic structure capable of stretching with the movements of respiration. (At full inspiration the bifurcation lies well below the manubriosternal joint.) The brachiocephalic and left common carotid arteries clasp the sides of the trachea (Fig. 4.3). The oesophagus lies behind the trachea.

The left principal bronchus passes under the arch of the aorta to reach the left lung in company with its pulmonary artery. The right principal bronchus passes under the azygos arch to get to the right lung. Remember this when we come to study the root of the lung shortly. Internally the bifurcation of the trachea is marked by an anteroposterior ridge called the **carina**. The right principal bronchus is more in line with the trachea, and is wider and shorter than the left principal bronchus, which travels more horizontally over the heart. Inhaled foreign bodies therefore tend to pass into the right main bronchus. The principal bronchi end at the point where they divide into **lobar bronchi**. On the right side this first division takes place outside the lung tissue. On the left, all the divisions of the bronchial tree lie within the lung tissue.

Many autonomic nerves mingle to form plexuses near the bifurcation of the trachea. We described earlier the **deep part** of the **cardiac plexus** that lies in front of the left principal bronchus and the **superficial part** of the **cardiac plexus** that lies around the ligamentum arteriosum under the arch of the aorta. Autonomic plexuses to the lungs, the **pulmonary plexuses**, are gatherings of autonomic nerves (parasympathetic and sympathetic) situated on the principal bronchi near both lung roots.

The bifurcation of the pulmonary trunk into right and left pulmonary arteries takes place in front of the left main principal bronchus (Fig. 4.6). Right and left arteries thereafter follow the principal bronchi to the lungs. Oxygenated blood is returned to the left atrium at the back of the heart by pulmonary veins. These lie below the bronchi and pulmonary arteries on either side.

## The roots of the lungs

Now is a good time to consider the roots of the lungs in more detail. We have described all of the structures that pass into, out of, and in front of the lung roots, and the pattern of structures at the hilum can be bet-

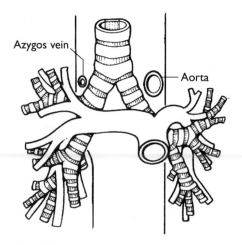

**Figure 4.6**   The left pulmonary artery runs over the top of and then behind the left main bronchus. The right pulmonary artery remains anterior to the right main bronchus.

ter explained now. It is a good time also to remind ourselves that there are many lymph nodes at the root of each lung. The upper part of the **left lung root** is occupied by the left pulmonary artery which also lies partly under the arch of the aorta close to the lung (Fig. 4.6). Note that in a fixed lung you can clearly see the position of the arch as a depression. The left pulmonary artery runs over the top and behind the left main bronchus. There are two pulmonary veins on the left and each is in contact with the pulmonary ligament below. The **right lung root** is similarly laid out but the comparison between the two lung roots is complicated by the fact that the right main bronchus divides before it enters lung tissue. Therefore, the upper lobe bronchus and its accompanying pulmonary artery are found together above the level of the main stem bronchus and its branch of the pulmonary artery. The right pulmonary artery is prevented from running up and over the right main bronchus by the upper lobe bronchus and remains anterior to it. Take another good look at Figure 4.6 and be sure you understand this arrangement. On the right, the arch of the azygos vein runs up and over the upper lobe bronchus and its accompanying branch of the pulmonary artery. The groove for this shows clearly on a fixed right lung above these two structures (Fig. 4.7). The two pulmonary veins on the right are disposed exactly as they are on the left, one in front of and one behind the

main bronchus and in contact with the pulmonary ligament. Since there is less of a difference between the pulmonary venous and arterial blood pressure than there is between the systemic arterial and venous blood pressure, it is not as easy to distinguish veins from arteries in the lung root by differences in the thickness of their walls. It is better to locate the bronchi by their thick cartilaginous walls and then recall the pattern of pulmonary vessels associated with them.

## The vagus nerves in the thorax

The two vagus nerves enter the thorax on either side of the trachea. They are destined to reach the oesophagus in the posterior mediastinum. On the right side the vagus lies in direct contact with the trachea (Fig. 4.8). It then passes behind the right lung root (bronchus, artery and vein) to reach the oesophagus. The left vagus is, as it were, held away from the trachea by the left common carotid artery and the arch of the aorta (Fig. 4.9). It also passes behind the root of the lung. Importantly, it gives off the recurrent laryngeal nerve on the left-hand side beneath the arch of the aorta. The right vagus nerve runs into the thorax on the side of the trachea. It gives off the right recurrent laryngeal nerve higher up than the left

**Figure 4.7**   The grooves created by the arch of the aorta and azygos vein are visible in the left and right fixed lungs respectively. In the root of the right lung (depending on where it is cut) the upper and lower lobe bronchi are each accompanied by a branch of the right pulmonary artery. The pulmonary veins lie low in the pulmonary ligament on either side of the lower lobe bronchus. In the left lung root the principal bronchus and accompanying pulmonary artery lie superiorly and the two pulmonary veins again lie lower, on either side of the bronchus.

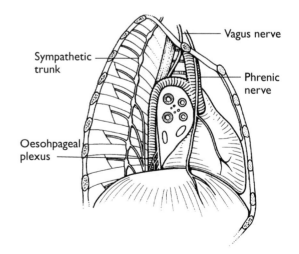

**Figure 4.8**   On the right the vagus nerve lies against the trachea in the superior mediastinum. It then passes behind the root of the lung and forms the oesophageal plexus around the oesophagus. The phrenic nerve passes in front of the lung root.

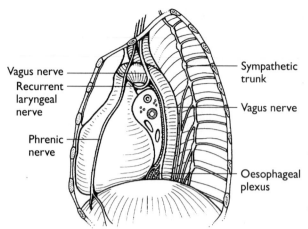

*Labels:* Vagus nerve, Recurrent laryngeal nerve, Phrenic nerve, Sympathetic trunk, Vagus nerve, Oesophageal plexus

**Figure 4.9**   The vagus nerve of the left passes on to the arch of the aorta where it gives off the left recurrent laryngeal nerve. It then runs on to the oesophagus behind the root of the lung.

vagus. The right recurrent laryngeal nerve hooks around the right subclavian artery and passes medially to gain the groove between the oesophagus and trachea.

Below the level of the lung roots, the vagi break up into a meshwork of nerves. These contribute to the oesophageal plexus surrounding the oesophagus (Fig. 4.10). The vagus is the sensory nerve to the whole oesophagus and secretor motor to the glands

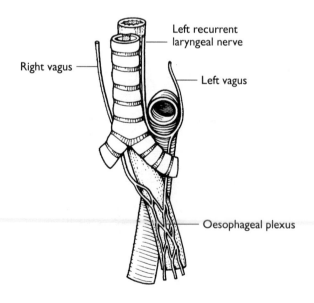

*Labels:* Right vagus, Left recurrent laryngeal nerve, Left vagus, Oesophageal plexus

**Figure 4.10**   The oesophageal plexus contains vagal nerve fibres which then reform as the right and left gastric nerves.

of the oesophageal mucous membrane. What remains of the vagal fibres reform as the right and left **gastric nerves** or **vagal trunks**. The gastric nerves contain fibres derived from both left and right vagus nerves. These, generally two on the right and one on the left, pass through the oesophageal opening of the diaphragm at the level of T10. The two right nerves pass to the coeliac plexus, which we will describe later, and the left nerve passes on to the anterior surface of the stomach.

## The thoracic aorta

The thoracic aorta is best divided, for descriptive purposes, into the **ascending aorta**, **arch of the aorta** and **descending aorta**. We have already described the great vessels that arise from the arch of the aorta. The concavity of the arch lies at the level of the junction between superior and middle mediastinum, that is at the level of the manubriosternal joint. The bulk of the arch therefore lies in the superior mediastinum. The arch is aligned in an almost anteroposterior direction. It arches over the right pulmonary artery and the left principal bronchus.

From the concavity of the arch, a remnant of a fetal vessel connects the aorta to the left pulmonary artery. This is the **ligamentum arteriosum**. It was originally the **ductus arteriosus**. During fetal life when the lungs do not function, blood is short-circuited through this channel from the pulmonary artery into the aorta. You will remember that the deoxygenated blood stream arriving in the right atrium of the fetus from the superior vena cava crosses the oxygenated blood stream. The deoxygenated stream passes through the tricuspid valve into the right ventricle and hence into the pulmonary trunk. There is no point in sending this stream through the functionless fetal lungs, and so it is short-circuited into the aorta through the ductus arteriosus. Thus the blood reaching the aorta is mixed, oxygenated blood coming from the left ventricle and deoxygenated blood coming through the ductus. The resultant level of oxygenation is, however, sufficient for the needs of the fetus.

The descending aorta starts behind the left principal bronchus. It descends on the left through the posterior mediastinum passing towards the midline (Fig. 4.11). It lies against the bodies of the thoracic vertebrae in its descent and grooves the outline of the

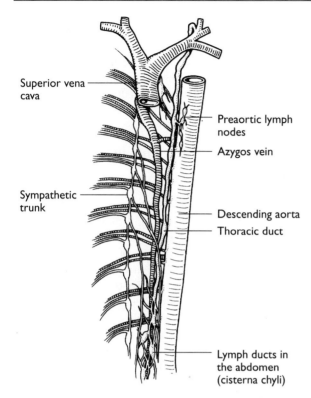

**Figure 4.11** The azygos vein, thoracic duct and descending aorta lie behind the oesophagus in the posterior mediastinum. The sympathetic trunks lie more laterally in the right and left paravertebral gutters.

Labels on figure:
- Superior vena cava
- Sympathetic trunk
- Preaortic lymph nodes
- Azygos vein
- Descending aorta
- Thoracic duct
- Lymph ducts in the abdomen (cisterna chyli)

vertebral bodies slightly on the left. We saw earlier that at each segmental level the aorta gives off posterior intercostal arteries. The aorta supplies some of the blood to the oesophagus by means of **oesophageal branches**, and blood to the tissues of the bronchi and lungs through **bronchial arteries**. The aorta eventually reaches the aortic opening in the diaphragm. From Figure 4.4 it can be seen that the oesophagus crosses over the aorta. To the right of the aorta lies the thoracic duct and the vena azygos.

# The oesophagus in the thorax

The oesophagus is a muscular tube that conveys food and liquid from the mouth to the stomach. The oesophagus enters the superior mediastinum in the midline through the superior aperture of the thorax. Here it lies behind the trachea. It does not lie in direct contact with the vertebral column here, since there is a small mass of prevertebral muscle covered with prevertebral fascia in this region. Lower down in the posterior mediastinum there are no prevertebral muscles. On either side, the edges of the lungs with their pleural covering reach the oesophagus.

The muscle of the oesophagus consists of an inner circular layer and an outer longitudinal layer. The upper third of the oesophagus contains voluntary muscle in its wall. This contracts rapidly to move food away from the airway and neck. Muscle in the lower two-thirds of the oesophagus is involuntary.

Behind the trachea, the oesophagus veers slightly to the left, so that by the time it reaches the posterior mediastinum it is lying behind the left principal bronchus. As it enters the posterior mediastinum it is crossed by the arch of the aorta on the left side and by the arch of the azygos vein on the right side. In its descent, however, the oesophagus moves to the left of the midline towards the oesophageal opening in the diaphragm. It therefore crosses in front of the descending aorta, as the aorta moves towards the midline.

The left atrium and the oblique sinus of the pericardial cavity lie immediately in front of the oesophagus below the level of the bifurcating trachea. In conditions where there is enlargement of the left atrium, such as in mitral stenosis, this chamber expands backwards and indents the oesophagus more than usual. Behind the oesophagus, and following the same oblique course as the aorta, is the **thoracic duct**. To the right of these structures is the ascending **vena azygos** channel.

There is an important constriction of the oesophagus; at its origin in the neck behind the larynx. There are three further important constrictions along the length of the oesophagus in the thorax. However, it is important to note that these are undetectable in an empty collapsed oesophagus they become clear only when a bolus of food is swallowed and can be studied on an oblique radiograph of the thorax during a 'barium swallow'. The first constriction in the thorax is where the left bronchus crosses the oesophagus, and the third is as it enters the stomach. These, together with the constriction in the neck behind the larynx, are the sites where swallowed food and drink are held up most in passage to the stomach. They are, perhaps also for this reason, common sites for carcinoma to develop. In addition to these, there is another constriction of the oesophagus in the thorax. An enlarged left atrium also presses into the oesophagus and its enlarged outline can be seen in an oblique radiograph of a barium swallow (Fig. 4.12). When, for example, a diseased mitral valve is unable to prevent blood from the left ventricle flowing back into the left atrium under pressure, it enlarges to cope with this.

Somatic motor nerves supply the voluntary muscle of the oesophagus, and parasympathetic motor nerves supply the involuntary muscles. These parasympathetic fibres run in the vagi and take part in the formation of the oesophageal plexus. Some sympathetic fibres also reach the oesophagus, but they are probably vasoconstrictor only in function. They intermingle with the parasympathetic fibres in the oesophageal plexus.

Most of the blood supply to the oesophagus is from branches of the aorta. However, in the superior mediastinum blood descends from the neck via the **inferior thyroid arteries**. Lower down, blood ascends from the abdomen via oesophageal branches of the **left gastric artery**. Venous return from the oesophagus follows a similar pattern and this, as we shall see, is clinically important. The middle parts drain into the azygos system and the upper part into the brachioce-

phalic veins of the neck. The lower veins drain through the oesophageal opening in the diaphragm into veins that accompany the left gastric artery. Lymphatic drainage follows the course of the arteries to lymph nodes in the neck, mediastinum and abdomen.

## The thoracic duct and lymph nodes in the mediastinum

Chains of lymph nodes in the thorax lie in front of the aorta (**preaortic nodes**) and alongside the aorta (**para-aortic nodes**). Other nodes lie along the internal thoracic artery on each side of the mediastinum. Around the trachea and its bifurcation there is a very important group of lymph nodes that drain the heart and lungs. These are the **tracheobronchial nodes**. This group drains into a mediastinal lymph trunk on each side of the thorax. The major chains drain upwards on their own or via the thoracic duct or right lymph duct towards the great veins in the superior mediastinum.

The **thoracic duct** enters the posterior mediastinum through the aortic opening. It carries lymph from the lower half of the body towards the veins in the root of the neck. It also receives lymph from the left side of the thorax. In its ascent, the thoracic duct follows the course of the aorta running from the midline to the left side. It therefore lies behind and eventually to the left of the oesophagus by the time it has reached the superior mediastinum.

In the neck the duct receives lymph from the left side of the head and from the left upper limb by means of **left jugular** and **left subclavian lymph trunks**. Eventually the thoracic duct enters one of the great veins on the left side of the neck, usually the brachiocephalic. The thoracic duct drains all lymph from the body except that from the right arm, the right side of the thorax and the right side of the head and neck. These are drained by the **right lymphatic trunk**, **right subclavian** and **right jugular trunks** respectively into the right brachiocephalic vein.

## The sympathetic trunks in the thorax

The **sympathetic trunks** extend from the neck to sacral region. They run in the right and left **paravertebral**

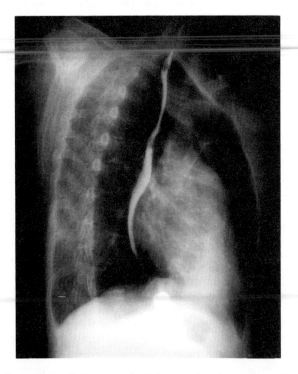

**Figure 4.12**    Radiograph of a barium swallow in a patient with an enlarged left atrium which is visible against the oesophagus. (After Pegington J. *Clinical Anatomy in Action*. Edinburgh: Churchill Livingstone, 1987.)

gutters on the necks of the ribs in the thorax (Figs 4.8 and 4.9). Below, each trunk passes into the abdomen beneath the medial arcuate ligament. Above they reach the neck by passing over the neck of the first rib. The sympathetic trunks in the thorax give off the **greater, lesser** and **least splanchnic nerves** (Fig. 4.13). For what its worth, the greater is derived from segments 5–9, the lesser from segments 10–11 and the least from segment 12. (They are then said to be like physicians, in that they arrive on the ward at 5 minutes to 9, have coffee at 10 minutes to 11 and then go home at 12.) These groups of preganglionic fibres pass through the thorax, pierce the diaphragmatic crura and carry preganglionic sympathetic fibres into the abdomen.

The vagus nerves and the sympathetic trunks carry efferent autonomic fibres to supply involuntary muscles and glands. In the thorax these include heart muscle (especially the SA node), glands and involuntary muscles of the lungs and bronchi, and glands and involuntary muscles of the oesophagus. The parasympathetic and sympathetic nerve fibres interweave with each other to form **autonomic plexuses**. There is, however, no functional communication between the two systems in these plexuses.

# Applied anatomy of some important structures in the superior and posterior mediastinum

There are several important conditions affecting the aorta. Occasionally, the ductus arteriosus does not close at birth. Since the pressure is then greater in the left ventricle and aorta, oxygenated blood escapes from the aorta into the pulmonary circulation. The condition is called a **patent ductus arteriosus** and requires surgery to tie it off. The aorta sometimes becomes dilated or constricted. A dilatation is called an **aneurysm** and a constriction is referred to as **coarctation of the aorta**. In the latter case, very little blood can get past the obstruction to supply the lower part of the body (Fig. 4.14). Collateral vessels then open up in an attempt to bypass the obstruction. For example, blood passing through the subclavian arteries can pass into the internal thoracic arteries and thereby the intercostal arteries. Once in the intercostal arteries, it can then re-enter the aorta below the level of a constriction in the posterior aortic arch.

Carcinoma of the lung spreads quickly through the

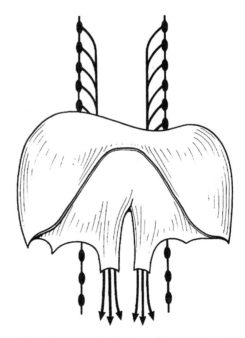

**Figure 4.13** The thoracic sympathetic splanchnic nerves arise from the sympathetic trunks. The greater, lesser and least splanchnic nerves pierce the crura of the diaphragm to enter the abdomen. The sympathetic trunks pass behind the medial arcuate ligaments to enter the abdomen.

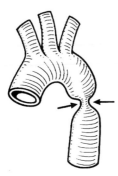

**Figure 4.14**   Coarctation of the aorta prevents or greatly restricts the passage of arterial blood to the lower part of the body.

lymph nodes of the lung root and chains in the mediastinum. Alterations in the voice that resemble hoarseness or whispering often result from cancerous lymph nodes invading the recurrent laryngeal nerve beneath the aortic arch. Pressure on the sympathetic trunk in the thorax, either at the neck of the first rib, which it crosses, or in the paravertebral gutters, can result in unopposed parasympathetic effects in the trunk and head and neck. These include hot flushed skin, absence of sweating (and a drooping eyelid and constricted pupil, which are explained further in Volume 3). This combination of symptoms is known as Horner's syndrome.

Occasionally, the lumen of the oesophagus remains undeveloped and ends in a blind pouch (**atresia**). To make things worse there is often an accompanying abnormal communication between the oesophagus and trachea called a **fistula**. Blood from the lower parts of the oesophagus drains through the oesophageal opening in the diaphragm into the left gastric veins around the stomach. Eventually, these drain through the liver to reach the inferior vena cava. In certain diseases of the liver (cirrhosis, for example) this venous passage through the liver is greatly reduced. Blood arriving at the liver is diverted back through the left gastric and oesophageal veins into the venous plexus of the oesophageal wall. From here the venous blood can get back to the heart via the azygos system and superior vena cava. However, the oesophageal plexus remains congested and is prone to severe haemorrhage through the mucosal wall and into the lumen of the oesophagus.

It is not uncommon for the oesophagus to ride up through the oesophageal opening of the diaphragm, carrying with it part of the stomach. This is called a **hiatus hernia**. The normal mechanisms for preventing regurgitation of stomach contents are lost. Regurgitation of food can lead to ulceration of the oesophageal mucosa, and chronic irritation and pain. Often, for this reason, carcinoma of the oesophagus is most common near its inferior end. Any patient complaining of difficulty swallowing and of pain in the shoulder should immediately be presumed to have carcinoma of the oesophagus until proved otherwise. Referred pain to the shoulder occurs because the phrenic nerve innervates the parietal peritoneum and parietal pleura as well as the pericardium abutting the diaphragm. Pain is then referred to that portion of the dermatomes of segments C3, C4 and C5 farthest away from the spinal cord. This includes the skin of the neck and the shoulder.

The oesophagus in the adult is about 25 cm long. The cricoid cartilage, which lies at the level of the upper limit of the oesophagus, is some 15 cm beyond the front teeth and the oesophageal opening into the stomach about 40 cm beyond. These are useful distances to know when passing an oesophagoscope through from the mouth to the stomach.

# Summary and Revision of the Thorax

Use the computed tomography (CT) scans of the thorax (Figs 5.1 and 5.2) to revise the relationships of the structures in the superior mediastinum. Remember, that when you look at a CT scan you are looking at the slice from below. In the first of the series, the air-filled trachea is prominent centrally. This indicates that the level of the section is above the bifurcation of the trachea (and therefore above the level of T4). You can also see a collapsed but still patent oesophagus posterior to the trachea. The three great arteries that arise from the arch of the aorta are clearly visible: the left subclavian, left common carotid and right brachiocephalic arteries. These are each labelled. The

brachiocephalic vein and azygos vein are easily seen on the right. In Figure 5.2 the section is just below the aortic arch. The ascending aorta and descending aorta are full of blood and are therefore radiodense and white on the film. The right and left bronchi are air filled and dark. The pulmonary trunk is running horizontally beneath the arch of the aorta from side to side but the superior vena cava is imaged well, next to the ascending aorta on the right. The window level settings in these films mean that they do not show any details of the lung fields, which therefore simply appear dark.

Now look at Figure 5.3 which is a posteroanterior chest radiograph. Identify the structures associated with the heart shadow beginning on the left with the

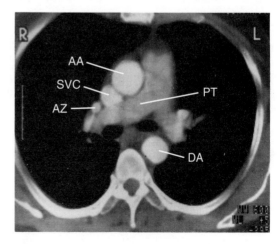

**Figure 5.1** Computed tomogram of the chest and superior mediastinum at the level of T3. Air in the trachea appears black. The lumen of the oesophagus is visible in front of the vertebral body. SA, left subclavian artery; CC, left common carotid artery; BA, brachiocephalic artery. The brachiocephalic vein (BV) and the azygos vein (AZ) are also well imaged.

**Figure 5.2** Computed tomogram of the chest and mediastinum at the level of T6. Air is visible in the right and left principal bronchi. The ascending aorta (AA) and descending aorta (DA), as well as the superior vena cava (SVC) and azygos vein (AZ) are each well imaged at this level. PT, pulmonary trunk.

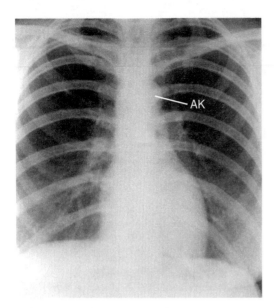

**Figure 5.3**   Posteroanterior chest radiograph. Move round the heart shadow from the aortic knuckle (AK) clockwise identifying the structures that form the border of the shadow. Use Figure 3.7 to check your identification.

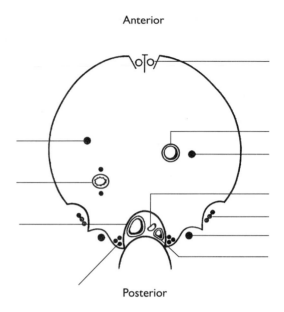

Anterior

Posterior

**Figure 5.4**   Diagrammatic representation of a transverse section through the diaphragm seen from below as in a computed tomogram. Identify the structures passing both through and behind the diaphragm. Use Figures 2.9 and 4.13 to do this if you have a problem.

aortic knuckle (AK) and moving round to the right clockwise. Refer back to Figure 3.7 to check your answers. Understand that blood in the pulmonary vessels (but not so much the bronchi, which are air filled) appears radiodense around the lung root and that the liver is similarly radiodense. Gas in the fundus of the stomach shows as a radiolucency under the diaphragm.

Figure 5.4 is a diagrammatic representation of the diaphragm seen in section. Identify each of the structures that passes through, in front of and behind the diaphragm. Refer back to the text in Chapter 2 and to Figure 2.9 to check your answers if you have problems with this.

Finally, to bring together what you have learned about the thorax, go though the multiple choice questions at the end of this chapter. For each stem, any one of the five answers (A)–(E) may be **either** correct or incorrect. There is no pattern to the correct combination of answers. Some of the questions are quite searching and on your first attempt at them you would be doing well to achieve a score of around 50% correct. We expect you to have to refer back to the text to discover the correct answers to some of the questions. This will improve your understanding of the anatomy of the thorax. On subsequent attempts at these multiple choice questions your score should improve.

# Multiple Choice Questions on the Peripheral Nervous System and Thorax

## 1. Sympathetic nerve fibres:
(A) arise only from segments T5 to T12 of the spinal cord
(B) innervate the SA node in the right atrium
(C) are responsible for bronchoconstriction
(D) only ever synapse in ganglia along the sympathetic trunk
(E) are responsible for vasodilatation of the skin

A ____ B ____ C ____ D ____ E ____

## 2. Parasympathetic nerve fibres:
(A) have a craniosacral outflow only
(B) generally synapse very close to the neuraxis
(C) are secretomotor to the mucous glands of the bronchi
(D) travel with the phrenic nerve in the thorax
(E) reach the limbs by running along blood vessels

A ____ B ____ C ____ D ____ E ____

## 3. In the peripheral nervous system:
(A) dorsal nerve roots contain both motor and sensory neurons
(B) each ventral ramus contains both motor and sensory neurons
(C) sensory neurons have their cell bodies in dorsal root ganglia
(D) postganglionic sympathetic nerve fibres are unmyelinated
(E) both parasympathetic and sympathetic autonomic ganglia contain synapses

A ____ B ____ C ____ D ____ E ____

## 4. The sternal angle:
(A) is part of the boundary of the inlet of the thorax
(B) is a secondary cartilaginous joint
(C) lies at the level at which the second costal cartilages meet the sternum
(D) is the joint between the sternal body and xiphisternum
(E) can be palpated in the living subject

A ____ B ____ C ____ D ____ E ____

## 5. The fourth intercostal nerves:
(A) give motor fibres to intercostal muscles
(B) convey parasympathetic neurons to the breasts
(C) contain fibres that are secretomotor to sweat glands
(D) are accompanied during part of their course by branches of the internal thoracic arteries
(E) are sensory to an area of skin on the anterior abdominal wall

A ____ B ____ C ____ D ____ E ____

## 6. Thoracic vertebrae:
(A) each articulate with four ribs (except at T1, T11 and T12)
(B) have articular facets on their transverse processes for articulation with the tubercles of ribs
(C) give origin to the right and left crura of the diaphragm
(D) are 12 in number
(E) may show flattening of the body where related to the descending aorta

A ____ B ____ C ____ D ____ E ____

## 7. The second intercostal nerve:
(A) has a collateral branch
(B) has an anterior cutaneous branch
(C) travels in the intercostal space above its corresponding artery and vein
(D) carries sympathetic fibres
(E) supplies the intercostal muscles of the second intercostal space

A ____ B ____ C ____ D ____ E ____

## 8. The diaphragm:
(A) contracts during expiration
(B) receives motor fibres from the vagus nerves
(C) forms a sphincter around the oesophagus
(D) arises in part as slips of muscle from the costal margin
(E) is firmly attached to the fibrous pericardium

A ____ B ____ C ____ D ____ E ____

## 9. The mediastinal surface of the right lung is related through the pleura to:
(A) the pericardium
(B) the vena azygos
(C) the right phrenic nerve
(D) the aortic arch
(E) the right recurrent laryngeal nerve

A ____ B ____ C ____ D ____ E ____

## 10. A stab wound through the 6th intercostal space into the lung tissue in the mid-clavicular line will pass through:
(A) the internal thoracic artery
(B) the costodiaphragmatic recess
(C) both internal and external intercostal muscles
(D) pectoralis major
(E) only one layer of pleura

A ____ B ____ C ____ D ____ E ____

## 11. The upper lobe of the right lung:
(A) is related through pleura to the trachea
(B) sometimes has a segment called the azygos lobe
(C) projects into the neck above the level of the clavicle
(D) is separated from the suprapleural membrane by both parietal and visceral pleura
(E) is related to the sympathetic trunk at the first rib

A ____ B ____ C ____ D ____ E ____

**12. The right principal bronchus:**
(A) has the azygos arch above it
(B) is more vertical than the left principal bronchus
(C) lies enclosed in the pulmonary ligament
(D) gives off the right upper lobe bronchus before entering the lung
(E) has cartilage in its walls

A＿＿ B＿＿ C＿＿ D＿＿ E＿＿

**13. The lateral arcuate ligament of the diaphragm:**
(A) bridges over the psoas muscle
(B) passes between the second lumbar vertebral body and its transverse process
(C) has the sympathetic trunk passing posterior to it
(D) lies posterior to the right kidney
(E) is related to the subcostal vessels

A＿＿ B＿＿ C＿＿ D＿＿ E＿＿

**14. The right lung:**
(A) has a section called the lingula
(B) has bronchopulmonary segments in each lobe
(C) has a transverse fissure
(D) is related (through pleura) to the right phrenic nerve
(E) is firmly fused with the central tendon of the diaphragm

A＿＿ B＿＿ C＿＿ D＿＿ E＿＿

**15. The root of the lung contains:**
(A) an artery or arteries that convey oxygenated blood
(B) parasympathetic fibres
(C) sympathetic fibres
(D) lymphatics and lymph nodes
(E) an artery or arteries that convey deoxygenated blood

A＿＿ B＿＿ C＿＿ D＿＿ E＿＿

**16. Features of the interior of the right atrium include:**
(A) musculi pectinati
(B) trabeculae carneae covering the wall of the auricular appendage
(C) a valve at the entrance of the superior vena cava
(D) the infundibulum
(E) a fossa ovalis

A＿＿ B＿＿ C＿＿ D＿＿ E＿＿

**17. The following structures can normally be traced along the margins of the mediastinal shadow on a posteroanterior radiograph of the chest:**
(A) the superior vena cava
(B) the aortic arch
(C) the interventricular groove
(D) the coronary sulcus
(E) the left auricular appendage

A＿＿ B＿＿ C＿＿ D＿＿ E＿＿

**18. The mitral valve:**
(A) possesses two major cusps
(B) 'guards' the right atrioventricular orifice
(C) has chordae tendineae
(D) is attached to the septomarginal trabecula (moderator band)
(E) closes during left ventricular contraction (systole)

A＿＿ B＿＿ C＿＿ D＿＿ E＿＿

**19. Chordae tendineae attach to:**
(A) cusps of the mitral valve
(B) papillary muscles
(C) cusps of the aortic valve
(D) the septomarginal trabecula (moderator band)
(E) cusps of the tricuspid valve

A＿＿ B＿＿ C＿＿ D＿＿ E＿＿

**20. The sinuatrial node:**
(A) lies in the wall of the right atrium near the superior vena cava
(B) imposes the rate of heart contraction
(C) usually receives its oxygenated blood from a branch of the left coronary artery
(D) receives sympathetic fibres
(E) receives parasympathetic fibres

A＿＿ B＿＿ C＿＿ D＿＿ E＿＿

**21. Branches of the right coronary artery may include:**
(A) the circumflex artery
(B) the anterior interventricular artery
(C) the marginal artery
(D) the sinuatrial (SA) artery
(E) vessels that supply the wall of the pulmonary trunk

A＿＿ B＿＿ C＿＿ D＿＿ E＿＿

**22. The following bound the oblique pericardial sinus:**
(A) the ascending aorta
(B) the inferior vena cava
(C) the right and left pulmonary veins
(D) the pulmonary trunk
(E) the left auricular appendage

A＿＿ B＿＿ C＿＿ D＿＿ E＿＿

**23. The oesophagus:**
(A) passes through the diaphragm at the level of T8
(B) is about 25 cm (10 in) in length
(C) receives some of its blood from inferior thyroid arteries
(D) drains some of its blood into the portal system
(E) is separated from the right atrium by the oblique sinus of pericardium

A＿＿ B＿＿ C＿＿ D＿＿ E＿＿

**24. The arch of the aorta is directly related to:**
(A) the left vagus nerve
(B) the ligamentum arteriosum
(C) the left recurrent laryngeal nerve
(D) the superficial cardiac plexus
(E) has its concavity at the level of the suprasternal (jugular) notch

A＿＿ B＿＿ C＿＿ D＿＿ E＿＿

**25. The oesophagus:**
(A) lies anterior to the thoracic duct
(B) has a normal constriction at the level of the bifurcation of the trachea
(C) has an oesophagogastric junction about 40 cm from the incisor teeth in adults
(D) can normally be seen on a posteroanterior chest film
(E) has a layer of voluntary muscle throughout its entire course

A＿＿ B＿＿ C＿＿ D＿＿ E＿＿

**26. The left phrenic nerve:**
(A) is a branch of the brachial plexus
(B) arises from cervical ventral rami
(C) lies against the trachea in the superior mediastinum
(D) passes behind the left lung root
(E) contains both motor and sensory fibres

A＿＿ B＿＿ C＿＿ D＿＿ E＿＿

### 27. Concerning the posterior origin of the diaphragm:

(A) the right crus arises from the bodies of T11 and T12
(B) some fibres of the right crus encircle the oesophagus
(C) the sympathetic trunk passes behind the lateral arcuate ligament
(D) the medial arcuate ligament is a thickening of the fascia covering the psoas
(E) the splanchnic nerves often pierce the crura

A＿＿ B ＿＿ C ＿＿ D ＿＿ E ＿＿

### 28. Structures passing through the oesophageal opening in the diaphragm include:

(A) vagal trunks
(B) the hemiazygos vein
(C) branches of the left gastric vessels
(D) lymphatics
(E) the left phrenic nerve

A＿＿ B ＿＿ C ＿＿ D ＿＿ E ＿＿

### 29. The thoracic descending aorta:

(A) passes behind the diaphragm at the level of the 12th thoracic vertebra
(B) gives off posterior intercostal arteries
(C) supplies branches to the oesophagus
(D) lies anterior to the oesophagus just above the diaphragm
(E) passes behind the diaphragm through a hiatus in company with the thoracic duct

A＿＿ B ＿＿ C ＿＿ D ＿＿ E ＿＿

### 30. The right vagus nerve:

(A) is found on the right surface of the trachea in the thorax
(B) passes in front of the right lung root
(C) contains nerve fibres whose activity speeds up the heart rate
(D) gives parasympathetic fibres to sweat glands
(E) contains fibres that take part in the formation of the oesophageal plexus

A＿＿ B ＿＿ C ＿＿ D ＿＿ E ＿＿

### 31. The thoracic duct:

(A) enters the thorax through the aortic opening of the diaphragm
(B) empties into the azygos vein
(C) is found behind the oesophagus in the thorax
(D) drains all lymph from both right and left sides of the thorax
(E) receives lymph from the pelvis

A＿＿ B ＿＿ C ＿＿ D ＿＿ E ＿＿

# Answers to Multiple Choice Questions

| | | | | | | | | | | | | | | | | | |
|---|---|---|---|---|---|---|---|---|---|---|---|---|---|---|---|---|---|
| 1. | AF | BT | CF | DF | EF | 12. | AT | BT | CF | DT | ET | 22. | AF | BT | CT | DF | EF |
| 2. | AT | BF | CT | DF | EF | 13. | AF | BF | CF | DT | ET | 23. | AF | BT | CT | DT | ET |
| 3. | AF | BT | CT | DT | ET | 14. | AF | BT | CT | DT | EF | 24. | AT | BT | CT | DT | EF |
| 4. | AF | BT | CT | DF | ET | 15. | AF | BT | CT | DT | ET | 25. | AT | BT | CT | DF | EF |
| 5. | AT | BF | CT | DT | EF | 16. | AT | BF | CF | DF | ET | 26. | AF | BT | CF | DF | ET |
| 6. | AT | BT | CF | DT | ET | 17. | AT | BT | CF | DF | ET | 27. | AF | BT | CF | DT | ET |
| 7. | AT | BT | CF | DT | ET | 18. | AT | BF | CT | DF | ET | 28. | AT | BF | CT | DT | EF |
| 8. | AF | BF | CT | DT | ET | 19. | AT | BT | CF | DF | ET | 29. | AT | BT | CT | DF | ET |
| 9. | AT | BT | CT | DF | EF | 20. | AT | BT | CF | DT | ET | 30. | AT | BF | CF | DF | ET |
| 10. | AF | BF | CT | DF | EF | 21. | AF | BF | CT | DT | ET | 31. | AT | BF | CT | DF | ET |
| 11. | AT | BT | CT | DT | ET | | | | | | | | | | | | |

# THE ABDOMEN

chapter
6

# The Anterior and Posterior Abdominal Walls

The abdominal cavity is separated from the thoracic cavity by the diaphragm above, but it is in continuity with the pelvic cavity below. The pelvic cavity is in turn limited by the pelvic diaphragm. The abdominal cavity is bounded by muscular walls at the front, sides and back. First of all in this chapter we will study the walls of the abdomen and then in subsequent chapters the contents of the abdominal cavity.

Classically, the anterior abdominal wall can be divided into different regions by two vertical and two horizontal lines for the purposes of physical examination and ease of description (Fig. 6.1). The **transpy-**

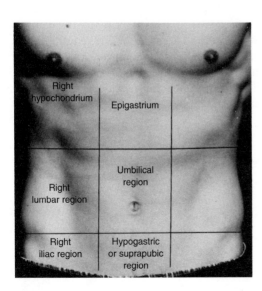

**Figure 6.1** Classically the anterior abdominal wall is divided into six regions by the transpyloric and transtubercular planes and the two vertical lines shown. In practice a simpler division into four quadrants is often used.

**loric plane** passes midway between the suprasternal notch and the pubic symphysis (or lower end of the body of the sternum and umbilicus). This plane lies at the level of L1. The umbilicus usually lies at a level between L3 and L4. The lower horizontal plane is called the **transtubercular plane** and passes through the tubercles of the iliac crests. The transtubercular plane lies at the level of L5. The **mid-inguinal point** is half way between the anterior superior iliac spine and the pubic symphysis. Lines drawn vertically up from the mid-inguinal points also correspond with the mid-clavicular lines. The mid-clavicular lines, the transpyloric plane and the transtubercular plane divide up the anterior abdominal wall into nine regions. These are, from the top down in the midline, the **epigastrium**, **umbilical** region and **hypogastric** (or **suprapubic** region), and, from the top down laterally, the right and left **hypochondrium**, **lumbar** regions and **inguinal** (or **iliac**). In practice a simpler division is often used, the anterior abdominal wall simply being divided into four quadrants.

Posteriorly, the abdominal cavity is bounded by the lumbar vertebral column and its associated prevertebral and postvertebral muscles. Laterally and anteriorly it is bounded by three layers of muscles that form the abdominal wall here. In the inguinal region of the male a structure called the **spermatic cord** transmits the duct (**ductus deferens**) vessels and lymphatics of the **testis** through the muscles of the anterior abdominal wall and onwards into the scrotum. In the female an embryological remnant follows this path through the anterior abdominal wall. These structures pass through what is known as the **inguinal canal** to get to the scrotum in the male or the labia majora in the female. The anatomy of this region and of the spermatic cord and inguinal canal will be studied here along with the muscles of the anterior abdominal wall. The anatomy of the inguinal region is important and it will be considered in some detail after a description of the abdominal wall musculature.

# The muscles of the anterior abdominal wall

The muscles of the anterior and lateral abdominal wall fill the space between the costal margin of the rib cage above and the iliac crest of the bony pelvis below. The most superior part of the bony pelvis runs around the iliac crests from the **posterior** to the **anterior superior iliac spines** (Fig. 6.2). In the midline anteriorly, the two pubic bones are joined by a secondary cartilaginous joint called the **pubic symphysis**. Just lateral to this is a crest of bone on the superior pubic ramus called the **pubic crest**. This ends as a raised tubercle an inch or two from the pubic symphysis, which is called the **pubic tubercle**.

The **rectus abdominis** muscle is a strap-like muscle that lies on each side of the midline. It runs from the pubic symphysis and pubic crest below to the margins of the costal cartilages just above the xyphoid process (Fig. 6.3). Three transverse tendinous intersections usually divide the muscle into four shorter muscle segments. One of these lies at the level of the umbilicus and another at the level of the xiphoid cartilage. A third tendinous intersection lies midway between the two. Rectus abdominis *flexes* the trunk, as when

rising from a bed. The **erector spinae** muscles are the principal agonists of the rectus abdominis muscles. They are powerful *extensor* muscles of the trunk. They lie between the spinous processes and the transverse processes of the vertebrae in the back and you can feel them easily as two powerful columns of muscle, each running just lateral to the midline of the back in the lumbar region.

The **external oblique** muscle is the most superficial muscle of the abdominal wall (Fig. 6.3). It takes origin from the lateral aspects of the lower ribs. Posteriorly, though, it has a free margin lateral to the erector spinae muscle (Fig. 6.4). Fibres of external oblique run downwards and forwards but become aponeurotic before reaching the rectus abdominis muscle anteriorly. Here, the aponeurotic part of external oblique helps in the formation of a sheath for the rectus abdominis. The aponeurosis of the external oblique is quite extensive inferiorly and no muscle fibres extend medially beyond a line joining the umbilicus and the anterior superior iliac spine. Inferiorly, the external oblique aponeurosis is firmly attached to the anterior superior iliac spine laterally and the pubic tubercle medially. Understand that it spans the gap between these two bony landmarks. The free lower border of the aponeurosis curls upwards and inwards on itself between these bony points and is known as the **inguinal ligament**. Through the gap beneath the free lower border of the inguinal ligament and the rim of the bony pelvis below it, there is a space for the femoral nerve and femoral vessels and the iliopsoas muscle to enter the thigh.

The **internal oblique** lies deep to the external oblique muscle (Figs 6.3 and 6.4). It arises from the

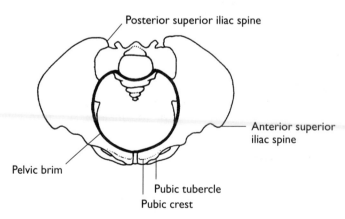

**Figure 6.2**    The iliac crests, pubic tubercles, pubic crest and pubic symphysis seen from above.

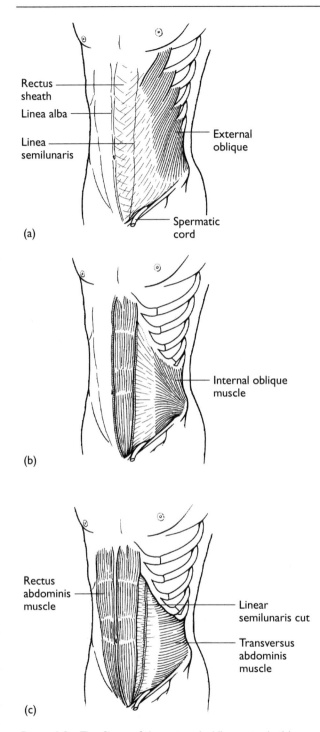

Rectus
sheath

Linea alba

Linea
semilunaris

External
oblique

Spermatic
cord

(a)

Internal oblique
muscle

(b)

Rectus
abdominis
muscle

Linear
semilunaris cut

Transversus
abdominis
muscle

(c)

**Figure 6.3** The fibres of the external oblique muscle (a) run downwards and forwards and become aponeurotic across the front of the rectus sheath. The fibres of the internal oblique (b) run in the opposite direction to these, and those of the transversus abdominis (c) run horizontally. (After Hall-Craggs ECB. *Anatomy as a Basis for Clinical Medicine*. Munich: Urban and Schwarzenberg, 1990.)

costal margin laterally but more posteriorly from tough fascia that encapsulates the erector spinae muscles between the spinous and transverse processes of the lumbar vertebrae. This fascia is called the **thoracolumbar fascia**. The fibres of the internal oblique muscle run upwards and forwards but, like the external oblique, become aponeurotic as they reach the rectus abdominis muscle anteriorly. Here they take part in the formation of the rectus sheath. Inferiorly, muscle fibres arise from the iliac crest and from the outer two-thirds of the curled up edge of the inguinal ligament. These lowermost fibres of internal oblique form an arch over the spermatic cord (see below) that first pass upwards and then over the spermatic cord and downwards into the bony crest of the pubis. As they approach the pubic crest they become aponeurotic here and fuse with deeper fibres of the aponeurosis of the transversus abdominis, which lies immediately behind. Together these two fused aponeurotic sheets form what is known as the **conjoint tendon**. The conjoint tendon is no more than a flat aponeurotic sheet that attaches to the pubic crest medial to the pubic tubercle. We will be discussing it again shortly.

The **transversus abdominis** muscle lies deep to the internal oblique (Figs 6.3 and 6.4). It also arises from the thoracolumbar fascia posteriorly, as well as from the inner margin of the lower ribs above and from the iliac crest and lateral part of the inguinal ligament below. Its fibres run horizontally, as its name implies but, like the external and internal oblique muscles, it becomes aponeurotic at the lateral margin of the rectus abdominis. Here the transversus abdominis also contributes to the formation of the rectus sheath. Below, as we have seen, the lowermost muscle fibres of the transversus abdominis also arch over the spermatic cord and then become aponeurotic as they pass down into the pubic crest to run into the conjoint tendon.

The muscles of the abdominal wall are lined on their deep surface with a fibrous membrane called the **transversalis fascia**. It is equivalent to endothoracic fascia in the thorax. The transversalis fascia varies in thickness and tends to be thinner in the more distensible regions of the abdominal wall. Beneath the transversalis fascia lies the **peritoneum**. Between the peritoneum and the transversalis fascia there is a layer of **extraperitoneal fat**.

The transversalis fascia runs across the posterior abdominal wall until it reaches the kidneys. At this

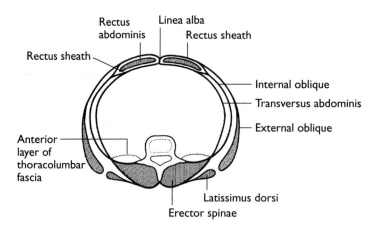

**Figure 6.4**   Posteriorly the external oblique has a free border. The rectus sheath is formed by the aponeuroses of the three muscles of the anterior abdominal wall. Thoracolumbar fascia encloses the erector spinae muscles as well as the quadratus lumborum on the posterior abdominal wall.

point it is quite thick and splits into two layers which pass in front of the kidneys and renal vessels, and behind them. These two layers are known as the **renal fascia** (they enclose a layer of **perirenal fat** between themselves and the capsule of the kidneys). Lower in the abdomen the transversalis fascia becomes the **fascia iliaca** and **psoas fascia**, which form tough layers in front of the iliacus and psoas muscles on the posterior abdominal wall.

## The rectus sheath

The rectus abdominis muscle is enclosed in a sheath (known as the **rectus sheath**), which is formed by the aponeuroses of the three anterolateral muscles of the anterior abdominal wall. The aponeurosis of the transversus abdominis muscle passes behind the rectus muscle. The aponeurosis of the external oblique muscle passes in front of the rectus muscle. The aponeurosis of the internal oblique muscle is said to split such that half of its fibres pass behind and half pass in front of the rectus abdominis muscle (Fig. 6.4). All these aponeurotic fibres fuse at a midline fibrous band called the **linea alba**. They also fuse at the lateral border of the rectus sheath along the so-called **linea semilunaris**. On a lean subject these lines of fusion can both be seen, together with the tendinous intersections of the rectus abdominis muscle. It is because the rectus sheath also fuses with the tendinous intersections on its anterior surface that they are easy to see.

At the midpoint between the umbilicus and the pubic symphysis there is a change in the relationship between the aponeurotic sheets of the anterior abdominal musculature and the rectus abdominis muscle. Below this point all of the aponeuroses pass in front of the rectus abdominis muscle. An **arcuate line** can be seen behind the muscle, indicating the level at which this occurs.

## Nerves and vessels of the anterior abdominal wall

The nerve supply to the anterior abdominal wall comes from the lower six intercostal nerves and the first lumbar nerve (T7 to L1 inclusive). These nerves run between the internal oblique and the transversus abdominis in the same way that the intercostal nerves run between the internal and innermost intercostal muscles (Fig. 6.5). However, a few of the lowermost nerves cannot gain entry into this neurovascular layer immediately, because they first appear on the anterior surface of large posterior abdominal wall muscles that get in the way. The intercostal and subcostal nerves pass round the abdominal wall, cross the linea semilunaris and enter the rectus sheath, and finally become cutaneous here. Nerves in the anterior abdominal wall extend along the course of the ribs and costal cartilages. The upper nerves run up towards the midline (T7 at the xiphoid cartilage) but the lower nerves run down towards the umbilicus (T10) or pubic symphysis (T12). L1 is the lowest

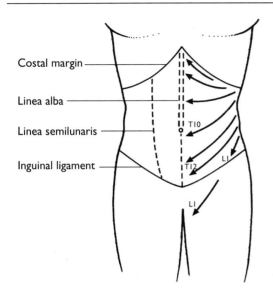

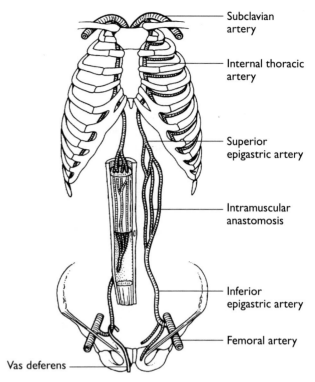

**Figure 6.5** Intercostal nerves run forwards into the anterior abdominal wall. The 7th and 8th run upwards, the 10th down to the umbilicus and the 12th down towards the pubic symphysis.

**Figure 6.6** An intramuscular anastomosis exists between the superior epigastric artery and the inferior epigastric artery within the rectus sheath. (After Green JH and Silver PHS. *An Introduction to Human Anatomy.* New York: Oxford Medical Publications, 1981.)

nerve to supply muscles of the anterior abdominal wall but it just fails to reach the midline.

The rectus sheath contains two important arteries. The **superior epigastric artery** is one of the terminal branches of the internal thoracic artery (remember, the other terminal branch of the internal thoracic artery is the **musculophrenic** artery). The superior epigastric artery enters the rectus sheath close to the xiphoid cartilage and runs down within it. The **inferior epigastric artery** is a branch of the external iliac artery (Fig. 6.6). The inferior epigastric artery first passes medially from the external iliac artery and hooks under the duct running from the testis, the ductus deferens, before turning upwards and entering the rectus sheath beneath the arcuate line. These two epigastric arteries run towards each other and anastomose freely within the rectus sheath. Branches from this anastomosis, and from the lower intercostal arteries and from lumbar branches of the abdominal aorta, supply the anterior and lateral abdominal wall.

A knowledge of the venous drainage of the anterior abdominal wall is also important. The superficial veins above the umbilicus drain to the internal thoracic vein and then to the subclavian veins and superior vena cava. The superficial veins below the umbilicus drain to the femoral veins and then to the inferior vena cava. Liver disease leading to obstruction of the portal vein can lead to distension of the epigastric veins since they offer an alternative route

for blood to bypass the obstruction and return to the heart via the superior vena cava. Distended veins of this sort around the umbilicus were said to resemble the head of Medusa, the character in Greek mythology whose head was writhing with snakes. Thus a so called '**caput Medusae**' around the umbilicus may indicate portal venous obstruction in the abdomen.

## Muscles of the posterior abdominal wall

The **psoas** muscles arise from either side of the lumbar vertebral column from the intervertebral discs and the adjacent parts of the bony vertebral bodies (Fig. 6.7). Tendinous arches pass from one attachment to the next, creating space for blood vessels to run behind the muscle and for nerves to escape through from the spinal canal. The psoas passes inferiorly and laterally

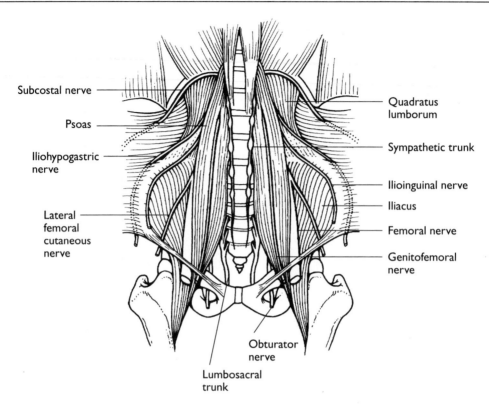

**Figure 6.7**   The subcostal nerve and the nerves of the lumbar plexus run on the posterior abdominal wall. They emerge from the psoas muscle. The sympathetic and lumbosacral trunks are also shown.

to leave the pelvis and enter the thigh under the inguinal ligament.

The **iliacus** muscle (Fig. 6.7) arises from the inner aspect of the iliac blade. Its fibres pass downwards and medially to meet those of psoas and there is a groove at their junction. The combined **iliopsoas muscle** enters the thigh under the lateral portion of the inguinal ligament and converges to a tendon which inserts into the lesser trochanter of the femur. Psoas and iliacus are powerful flexors of the thigh. Psoas and iliacus get their nerve supply from segments L2–L3.

The **quadratus lumborum** muscle (Fig. 6.7) forms part of the posterior abdominal wall just lateral to the psoas muscle. Its fibres arise from the lower aspect of the 12th rib and from the tips of the transverse processes of the lumbar vertebrae (which are, after all, rib elements of the vertebrae that never fully developed). Its fibres pass downwards and laterally to insert on to the posterior aspect of the iliac crest. Quadratus lumborum is a lateral flexor of the vertebral column in the lumbar region. A thin layer of the tough thoracolumbar fascia that surrounds the erector spinae muscles continues anteriorly to run

over the surface of quadratus lumborum (Fig. 6.4). Quadratus lumborum gets its nerve supply directly from the adjacent lumbar nerves L1 to L4.

So that the psoas and quadratus lumborum muscles can move independently from the muscle fibres of the diaphragm, which run over their anterior surface as far down as L1 and L2, two fibrous arches span across them. These are the **lateral** and **medial arcuate ligaments** and they are part of the posterior inferior margin of the diaphragm. In addition there are two muscle slips that rise into the diaphragm from the vertebral bodies near the midline. These come from the bodies of L1 and L2 on the left and from L1, L2 and L3 on the right. The muscle slips are called the **crura**. The **right crus** rises up into the diaphragm and curls around the oesophageal opening at the level of T10 and contributes to the sphincteric action of the oesophagus here. The **left crus** passes upwards and its fibres simply mingle with other muscular slips of the diaphragm arising from the arcuate lines and lower ribs. Across the midline and in front of the opening for the aorta, which passes behind the diaphragm at the level of T12, the crura are bound to each other by another fibrous band, called the **median arcuate ligament**.

# Nerves of the posterior abdominal wall

We have already seen how the intercostal nerves of spaces T7 to T11 (and the subcostal nerve T12), together with the anterior primary ramus of L1, innervate all the skin and the muscles of the anterior abdominal wall. The ventral rami of L1 to L4 take part in the **lumbar plexus**, a mixing of nerves within the substance of the psoas muscle. It follows that each of the nerves that run across the posterior abdominal wall first emerges from the psoas muscle. The most superior nerve to run across the posterior wall of the abdomen is the **subcostal nerve**.

The subcostal nerve (Fig. 6.7) arises behind the diaphragm just below the 12th rib. (Understand that this in fact means low in the thoracic cavity.) It then passes beneath the lateral arcuate ligament on to the anterior surface of the quadratus lumborum. Here it lies behind the kidney. The subcostal nerve then passes laterally and runs off the quadratus lumborum muscle. It is now able to pierce the transversus abdominis muscle and run in the neurovascular plane. It travels round in this plane to enter the rectus sheath and become cutaneous just above the pubic symphysis.

The first lumbar ventral ramus gives off a collateral branch, exactly like an intercostal nerve, while still within the psoas muscle. The main trunk and the collateral branch have separate names. These are the **iliohypogastric nerve** and the **ilioinguinal nerve** respectively. The iliohypogastric nerve passes out of the psoas muscle and on to the anterior surface of the quadratus lumborum (Fig. 6.7). It passes downwards and laterally to run off this muscle just above the iliac crest, where it can now pierce the transversus abdominis muscle and continue in the neurovascular plane. The iliohypogastric nerve does not end up supplying the rectus abdominis muscle. In the anterior abdominal wall it pierces the internal oblique muscle as well, and finally also the external oblique aponeurosis to supply the skin just superior to the pubic symphysis.

The ilioinguinal nerve also runs out of the psoas muscle on to the anterior surface of the quadratus lumborum muscle below the level of the iliohypo-

gastric nerve (Fig. 6.7). Both nerves, like the subcostal nerve, run behind the kidney here. However, as it runs further laterally, the ilioinguinal nerve passes on to the iliacus muscle such that it still cannot get into the neurovascular plane until it reaches the transversus abdominis muscle at the anterior superior iliac spine (where this muscle arises at this level). When in the neurovascular plane it supplies the lower fibres of the transversus abdominis and internal oblique muscle fibres. Like its main trunk partner (the iliohypogastric nerve) it also quickly pierces the internal oblique muscle to lie between this muscle and the external oblique aponeurosis. In this plane it enters the inguinal canal and then leaves it at the so-called **superficial ring**, piercing the external spermatic fascia to become cutaneous here. It is sensory to the root of the penis, to the anterior aspect of the scrotum and to the adjacent part of the thigh. Its course in the inguinal region will be covered again shortly.

The **genitofemoral nerve** is made up of nerve fibres from L1 and L2 (Fig. 6.7). The nerve runs out of the psoas muscle on its anterior surface. It continues to run down its anterior surface and then on to the lateral aspect of the external iliac artery as it travels towards the inguinal ligament. At the inguinal ligament the nerve divides and the **genital branch** passes into the deep inguinal ring to supply the **cremaster muscle**. The **femoral branch** passes under the inguinal ligament on the front of what is now the femoral artery and pierces the fascia of the thigh to supply the skin on the front of the thigh below the inguinal ligament.

The **lateral femoral cutaneous nerve of the thigh** is composed of nerve fibres from L2 and L3. It passes out of psoas and directly on to the iliacus muscle. At the anterior superior iliac spine it passes out into the thigh beneath the inguinal ligament to supply the skin of the lateral aspect of the thigh.

Two other important nerves are formed in the lumbar plexus which pass down the posterior wall of the abdomen to innervate muscles of the lower limb. These are the **femoral** and **obturator nerves** (Fig. 6.7). Both are formed from fibres whose segmental origin is L2 to L4 but the femoral nerve forms from the dorsal divisions and the obturator from the anterior divisions of the ventral rami of these nerve segments. The femoral nerve runs in the groove between the iliacus and psoas muscles and passes out into the thigh under the inguinal ligament. The obturator nerve appears at the medial border of the psoas muscle at the level of the sacroiliac joint and then runs along the lateral wall of the pelvis, lateral to the internal iliac artery and vein and the ureter. It leaves the pelvis through the obturator foramen just beneath the superior pubic ramus.

## The inguinal canal

First look carefully again at an articulated pelvis, if you can, and identify the anterior superior iliac spine, pubic tubercle, pubic crest as before and also now the **pectineal line**. The inguinal ligament runs between the anterior superior iliac spine and the pubic tubercle (Fig. 6.8). Identify the pubic crest on to which the rectus abdominis muscle attaches and also the conjoint tendon a bit further lateral to the rectus abdominis. The pectineal line of the pubic bone runs horizontally and laterally from the crest to become continuous with the margin or brim of the true pelvis. Fibres of the inguinal ligament sweep backwards and horizontally from the ligament and pubic tubercle to the pectineal line, to form the triangular **lacunar ligament**. The free lateral margin of the lacunar ligament is crescent shaped and is the medial boundary of the femoral canal beneath the inguinal ligament. The details of this region will be studied with the lower limb. When the pelvis is correctly oriented, the upper surface of the lacunar ligament lies horizontally and acts as a floor to support the spermatic cord here as

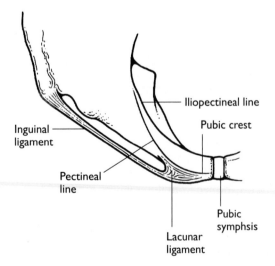

**Figure 6.8**  As the inguinal ligament approaches the pubic tubercle, the lacunar ligament runs from it on to the pectineal part of the iliopectineal line.

it passes out into the scrotum (Fig. 6.8). The lacunar ligament is continued as a band, the **pectineal ligament**, along the pectineal line.

## Descent of the testis

The gonads first develop high in the abdominal cavity in the region of the suprarenal glands. They are retroperitoneal structures which migrate down to the pelvis (in the case of the ovary) and onwards through the inguinal canal and into the scrotum (in the case of the testis). They bring their blood supply with them from the aorta such that the gonadal arteries reflect the path of their migration. A fibrous structure known as the **gubernaculum** (or 'helmsman', a term coined by John Hunter 1728–1783) may guide or even pull the gonad along a path of least resistance during this migration. A remnant of the gubernaculum in the form of the ligament of the ovary and the round ligament of the uterus persists into adulthood in females. In the pelvis it runs through the inguinal canal and into the labia majora in females.

As the gubernaculum draws the testis downwards it adheres to a portion of the peritoneum at the entrance to the inguinal canal (Fig. 6.9). A pouch of peritoneum therefore precedes the testis through the anterior abdominal wall as it is pulled down into the scrotum. The pouch is known as the **vaginal process** and it elongates and narrows in its middle portion. A balloon of peritoneum eventually buds off at the distal end of the vaginal process in the scrotum. The rest of the elongated neck of the pouch degenerates and usually disappears. The testis, being retroperitoneal, passes through the anterior abdominal wall *behind* the vaginal process and comes to press into the back of the balloon in the scrotum. In adult men the testis is still covered on its anterior surface and its sides by a double-layered serous lined sac, the **tunica vaginalis**. This allows frictionless movement of the testis within the scrotum. After an injury or through disease, the tunica vaginalis may fill with fluid and is then known as a **hydrocele**.

## The spermatic cord

The testis carries the layers of the anterior abdominal wall with it as it passes out into the scrotum. It also

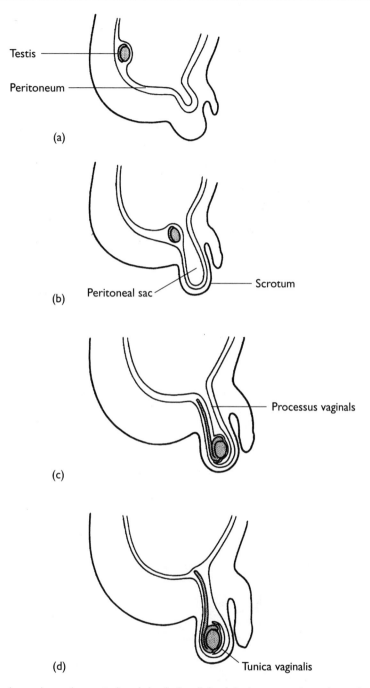

Testis

Peritoneum

(a)

Scrotum

Peritoneal sac

(b)

Processus vaginals

(c)

Tunica vaginalis

(d)

**Figure 6.9** As the testis descends on the posterior abdominal wall (a–c) it draws a peritoneal sac along with it (the processus vaginalis). This usually buds off and forms the tunica vaginalis in the adult (d). (After Hall-Craggs ECB. *Anatomy as a Basis for Clinical Medicine*. Munich: Urban and Schwarzenberg, 1990.)

draws its blood supply (the testicular artery) and duct (the vas deferens), as well as its nerve supply and lymphatics, with it through the anterior abdominal wall. These centrally situated structures become covered with sleeves of fascia which are each derived from the layers of the anterior abdominal wall. These all constitute the spermatic cord.

Three fascial sleeves of the spermatic cord can be

recognized in the adult (Fig. 6.10). The innermost layer is an extension of the transversalis fascia and is known as the **internal spermatic fascia**. A muscular layer intertwined with loose fascia is derived from the internal oblique muscle and is known as the **cremasteric muscle and fascia**. (Since the testis passes *below* the lowermost muscle fibres of the transversus abdominis it fails to draw this muscle onwards with it in the form of a separate sheath, although some people argue that it probably forms a small part of the cremasteric fascia.) A tough layer on the outside of the spermatic cord, called the **external spermatic fascia**, is derived from the external oblique aponeurosis.

The cord contains the vas deferens and three arteries: one to the vas deferens from the vesical artery to the bladder, one to the cremaster muscle from the inferior epigastric artery, and the testicular artery to the testis from the aorta.

The cord also contains a plexus of veins called the **pampiniform plexus** as well as nerves and lymphatics. The pampiniform plexus may act as a heat exchanger to cool the arterial blood destined for the testis and so help to keep it at the optimum temperature for spermatogenesis. The venous plexus reduces to three or so veins at the superficial ring, which then drain as one vein into the inferior vena cava on the right or into the renal vein on the left. These are the testicular veins.

Three nerves are associated with the spermatic cord. The genital branch of the genitofemoral nerve enters the deep ring and passes through the inguinal canal and supplies the cremaster muscle. Its action can be tested by eliciting the **cremasteric reflex**: stroking the inner aspect of the thigh causes the testis to rise as the cremaster muscle contracts. The genital branch of the genitofemoral nerve probably also carries sympathetic fibres which travel onwards to the **dartos muscle** in the scrotum. These nerves have travelled as far as the spermatic cord along blood vessels. The dartos muscle is smooth muscle which is probably also involved in temperature regulation of the testis by relaxing and contracting the skin of the scrotum in response to heat and cold. The ilioinguinal nerve runs over the iliacus muscle and pierces the transversus abdominis at the anterior superior iliac spine to enter the neurovascular plane. It then also pierces the internal oblique muscle and runs in the inguinal canal between this muscle and the external oblique aponeurosis. It emerges at the superficial ring between the same layers on the underside of the spermatic cord. Here it leaves the cord by piercing the external spermatic fascia and is the sensory nerve to

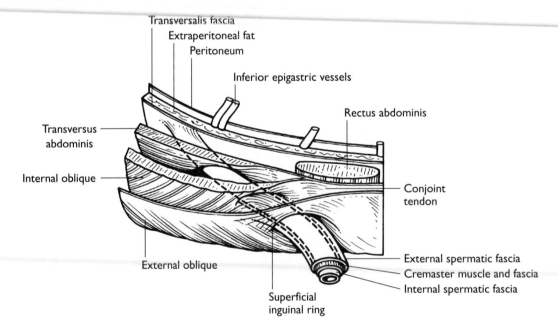

**Figure 6.10**    The contents of the spermatic cord pass through the layers of the anterior abdominal wall from the deep ring in the transversalis fascia through to the superficial ring in the external oblique aponeurosis. The external, cremasteric and internal spermatic fascia that cover the cord are derived from these layers. (After Yeager VL (1992) Intermediate inguinal ring. *Clinical Anatomy* **5**: 289–295.)

the root of the penis, the inner aspect of the thigh and the skin of the scrotum anteriorly. Lymphatics from the scrotum drain to the inguinal nodes but it is very important to remember that lymph from the testis drains through the spermatic cord to the **para-aortic lymph nodes**.

In summary then, as well as the vas deferens and lymphatics there are at the level of the superficial inguinal ring three arteries, three veins and three nerves contained within the three layers of fascia of the spermatic cord.

# The testis

The testis is suspended in the scrotum by the spermatic cord. The three fascial coverings of the spermatic cord are continued into the wall of the scrotum so that they cover the testis and tunica vaginalis. The tunica vaginalis is one of the layers covering the testis (Fig. 6.11). Deep to the tunica vaginalis is the tough **tunica albuginea**, which is the actual fibrous covering of the testis, and this in turn covers an innermost layer the **tunica vasculosa**. The head of the epididymis

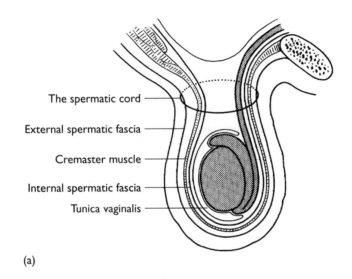

The spermatic cord
External spermatic fascia
Cremaster muscle
Internal spermatic fascia
Tunica vaginalis

(a)

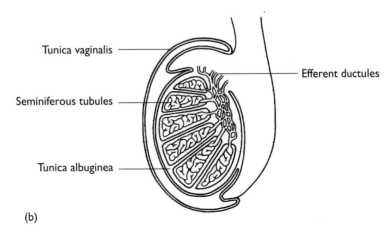

Tunica vaginalis
Seminiferous tubules
Tunica albuginea
Efferent ductules

(b)

**Figure 6.11** The layers of the spermatic cord surround each testis within the scrotum (a). The tunica vaginalis covers its anterior surface. Seminiferous tubules empty into efferent ductules (b) which form the head of the epididymis. (After Hall-Craggs ECB. *Anatomy as a Basis for Clinical Medicine*. Munich: Urban and Schwarzenberg, 1990.)

forms from the efferent ductules that pass from the lobules of the testis to the upper pole of the testis on its posterior surface (Fig. 6.12). The epididymis is an extremely long thin duct wound into a head, body and tail portion and which becomes continuous with the larger vas deferens at the inferior pole of the testis posteriorly. The tunica vaginalis presses in between the epididymis and the testis on the lateral side to form the sinus of the epididymis. The ductus deferens feels hard (like 'whip-cord') and initially runs up the back of the epididymis towards the upper pole of the testis (Fig. 6.12), and then back through the spermatic cord to the side wall of the pelvic cavity and finally across its floor to reach the prostate gland.

## Applied anatomy of the inguinal region

At birth the testis has usually already descended into the scrotum. At this time the path of the testis through the fascia transversalis, transversus abdominis, internal oblique and external oblique muscles is directly aligned front to back just lateral to the pubic tubercle. Gradually, during growth, the opening in the transversalis fascia (or deep ring) moves laterally nearer to the anterior superior iliac spine. The opening through the external oblique aponeurosis (or superficial ring) remains medially placed just next to the pubic tubercle. The path of the spermatic cord under the lowermost fibres of the transversus abdom-

inis muscle moves to an intermediate position between the deep and superficial rings. In this way the inguinal canal comes to run obliquely from lateral to medial through the anterior abdominal wall (Fig. 6.10).

This arrangement provides some strength to a potentially weak area of the anterior abdominal wall and resists displacement of structures from the abdominal cavity into and through the inguinal canal. The **floor** of the inguinal canal is the gutter-like horizontal surface of the lacunar ligament. It should be clear now (Fig. 6.10) that, lateral to the superficial ring, the external oblique aponeurosis and part of the origin of the internal oblique muscle lie in front of the spermatic cord and form the **anterior wall** of the inguinal canal. The transversalis fascia and the conjoint tendon lie behind the spermatic cord medial to the deep ring and form the **posterior wall** of the inguinal canal. The arching fibres of the internal oblique muscle form a **roof** over the inguinal canal.

Hernias in the inguinal region are common (Fig. 6.13). When the peritoneum lining the abdominal cavity enters the deep ring and protrudes through the inguinal canal, out through the superficial ring and into the scrotum, it is known as an **indirect inguinal hernia**. This type of hernia, even though it may present first in adults, is considered congenital. The sac of the hernia may contain a loop of bowel which can strangulate at the tight deep ring. The more usual hernia of old age is known as a **direct inguinal hernia**.

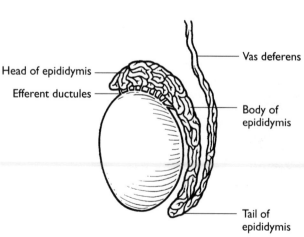

**Figure 6.12**  At the inferior pole of the testis, the tail of the epididymis becomes continuous with the vas deferens, which then runs up within the spermatic cord.

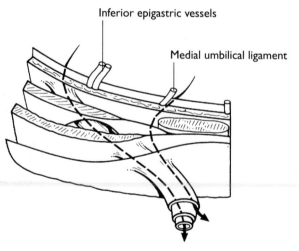

**Figure 6.13**  At operation an indirect inguinal hernia is found to pass lateral to the inferior epigastric vessels and a direct hernia medial to the vessels. (After Yeager VL (1992) Intermediate inguinal ring. *Clinical Anatomy* **5**: 289–295.)

Here, due to weakness and thinning of the conjoint tendon (which forms the posterior wall of the canal medially), the hernia can bulge through the conjoint tendon and then directly through the superficial ring and again into the scrotum. Identifying the pubic tubercle and the position of the inferior epigastric artery allows the origin of each type of hernia to be diagnosed. At operation, an indirect hernia passes lateral to the vessel and a direct hernia medial to the inferior epigastric artery.

# Retroperitoneal Structures and Development of the Mesenteries

In this chapter we will look first at the great vessels of the abdomen and their relationships to each other. Then we will look at the kidneys and suprarenal glands, which are retroperitoneal structures lying on the posterior abdominal wall. Next we will describe the mesenteries and ligaments of the abdominal cavity and try to relate their adult form to the developmental rotations of the foregut and midgut.

## The abdominal aorta

The aorta enters the abdomen by passing through the **aortic hiatus** behind the diaphragm at the level of T12. You will remember that this hiatus is behind the *median* arcuate ligament that connects the two crura. The aorta descends along the left anterior surface of the lumbar vertebral bodies until it bifurcates to form the two common iliac arteries at the lower border of

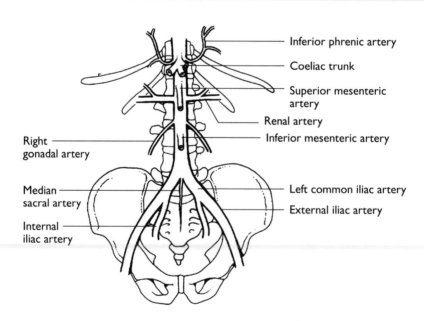

*Figure 7.1*   The abdominal aorta and its major paired and midline branches.

L4 (Fig. 7.1). This is roughly just below the level of the umbilicus in most people. Each common iliac artery then gives off an internal iliac artery at the level of the lower border of L5. The internal iliac artery runs into the pelvis. The common iliac artery continues as the external iliac artery and passes under the mid-inguinal point, midway between the anterior superior iliac spine and the symphysis pubis, to enter the thigh as the femoral artery.

The abdominal aorta gives off four paired lumbar branches that immediately run behind the psoas (behind its tendinous arches) and quadratus lumborum muscles. These quickly gain entry to the neurovascular plane between the transversus abdominis and internal oblique muscles. These arteries supply the muscles and skin of the abdominal wall in exactly the same way as the intercostal arteries do. A **median sacral artery** (which really represents the fifth pair of lumbar arteries fused together) arises at the bifurcation of the aorta and runs down into the pelvis to supply the muscles on the posterior pelvic wall.

Besides the lumbar arteries there are four other paired arteries that arise from the abdominal aorta. The first pair are the **inferior phrenic arteries** that supply the diaphragm. The second pair are the middle **suprarenal arteries** and the third pair are the **renal arteries**. The fourth pair we have already mentioned in our study of the inguinal region and these are the **gonadal arteries** (testicular or ovarian).

It remains only to describe the three unpaired midline branches of the abdominal aorta that supply blood to the gut and its derivatives. The **coeliac trunk** and the **superior mesenteric artery** arise from the anterior midline of the aorta high in the abdomen in front of the body of the first lumbar vertebral body. These two arteries supply the gut and its associated structures as far down as the last part of the transverse colon. The **inferior mesenteric artery** is the third midline branch that arises lower down on the anterior surface of the aorta at the level of L3. The inferior mesenteric artery supplies the distal part of the transverse colon and the descending colon, and part of the rectum and anal canal.

## The inferior vena cava and the veins of the abdomen

Understanding the relationship between the aorta and the inferior vena cava helps to explain several ana-tomical findings in the abdomen. Each of the arteries we have just described has an accompanying vein. However, the lumbar veins drain into the abdominal parts of the azygos and hemiazygos veins, which are called the **ascending lumbar veins** in the abdomen. These run upwards into the thorax deep to the psoas muscles in the abdomen. They do, none the less, have important anastomotic connections with other large veins in the abdomen.

Although blood passes up the inferior vena cava it may help when describing the inferior vena cava to follow its course from the diaphragm to the pelvis in the same way we described the abdominal aorta. The inferior vena cava passes through the diaphragm in its central tendinous portion at the level of T8. It immediately passes through the liver and emerges to the right of the abdominal aorta with the right crus lying between the two. The inferior vena cava stays on the right side of the vertebral bodies of the lumbar vertebrae, applied to the right side of the abdominal aorta as it descends through the abdomen (Fig. 7.2). At its bifurcation the inferior vena cava passes behind the right common iliac artery. From here on, the right and left common iliac veins, and later the internal and external iliac veins, travel on the inferomedial surfaces of the corresponding iliac arteries. The external iliac veins therefore enter the thigh as the femoral veins on the medial side of the femoral arteries.

Blood from the gut does not return to the inferior vena cava but to the **portal vein** and liver. We have seen already that the paired lumbar veins do not drain into the inferior vena cava either. Moreover, because the inferior vena cava lies on the right of the abdominal cavity, certain veins that 'should' drain into it on the left empty into the left renal vein instead. (Even the median sacral vein drains into the left common iliac vein and not into the bifurcation of the inferior vena cava because it lies too far over to the right.) All this means that there are fewer tributaries of the inferior vena cava than there are branches of the aorta.

The tributaries of the inferior vena cava are the two or three **hepatic veins** that arise and drain into the inferior vena cava within the substance of the liver, the right and left **renal veins**, but only the right **inferior phrenic**, right **suprarenal vein** and right **gonadal vein**. The left suprarenal vein and left gonadal vein drain directly into the left renal vein. (The left inferior phrenic vein drains into the left suprarenal vein.) It will help you consolidate these relationships

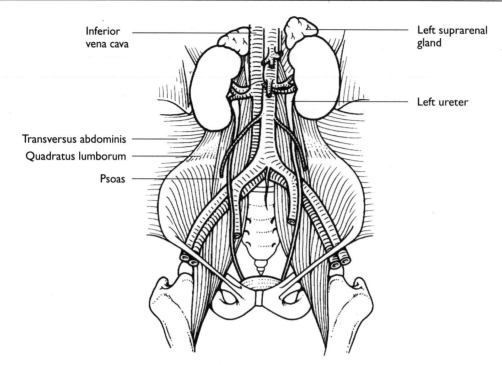

Inferior
vena cava

Left suprarenal
gland

Left ureter

Transversus abdominis
Quadratus lumborum
Psoas

**Figure 7.2**   The muscles of the posterior abdominal wall include the psoas, quadratus lumborum, transversus abdominis and the diaphragm. The kidneys lie against these muscles. The inferior vena cava lies to the right of the abdominal aorta but then bifurcates behind the right common iliac artery.

in your mind if you think for a moment about how the aorta and inferior vena cava pass through the diaphragm. Since the inferior vena cava pierces the diaphagm anterior to the aorta, it is easy to remember that the renal veins must then lie in front of the renal arteries as they run into the inferior vena cava.

## The kidneys, ureters and renal vessels

The kidneys lie encapsulated in their two layers of renal fascia opposite the first, second and third lumbar vertebral bodies (although be aware that they do move a bit during respiration and on moving about). The kidney itself is covered with a thin capsule and the perirenal fat lies between this and the renal fascia. Posteriorly, between the posterior layer of renal fascia and the fascia transversalis, there is another layer of fat called the **pararenal fat**. The hilum of the kidney lies roughly in the transpyloric plane, with the left kidney being a bit above this level and the right a bit below.

Behind the kidneys lie the psoas muscle, the quadratus lumborum muscle and the transversus abdominis muscle further laterally (Fig. 7.2). Both kidneys also lie over the lower edge of the diaphragm and so are related through it to the pleura and costodiaphragmatic recess posteriorly. The subcostal, iliohypogastric and ilioinguinal nerves run across the posterior abdominal wall between these muscles and the kidneys.

The anterior layer of renal fascia runs across the renal veins and arteries to merge with the connective tissue covering the aorta and inferior vena cava. Later on we shall see that, anteriorly, the right kidney is directly related to the second part of the duodenum, gall bladder and the hepatic flexure of the colon through the anterior layer of renal fascia. The small intestine and the liver also relate to the anterior surface of the right kidney through their peritoneal coverings. The left kidney is directly related to the body of the pancreas and to the splenic flexure of the colon through its anterior layer of renal fascia. The spleen, stomach and small intestine also relate to the left kidney but through their peritoneal coverings.

The kidneys begin to develop in the pelvis and then

rise up the posterior abdominal wall bringing the ureters with them. The kidneys get their early blood supply from lower down in the abdomen but this almost always disappears to be replaced by the renal arteries and renal veins higher up.

The ureters run down to the bladder on the surface of the psoas muscles. On an intravenous urogram they can be seen to run down across the tips of the lumbar transverse processes and then across the bifurcation of the common iliac arteries and into the pelvis (Fig. 7.3). The ureters pick up their blood supply from different sources as they descend through the abdomen. The renal arteries, the aorta and common iliac arteries all supply the ureters in the abdomen. The ureters are an important posterior relation of the gonadal vessels and duodenum on the right. They run over the common iliac vessels and often this is where kidney stones lodge.

The ureters enter the hilum behind the renal arteries (which divide into several branches at the hilum). The renal arteries, as we have already seen, lie posterior to the renal veins. The ureter opens out in the kidney to become the **renal pelvis** which, together with the larger vessels here and some fat, is known as the **renal sinus** (Fig. 7.4). The renal pelvis bifurcates or trifurcates to form the **major calyces**. In turn each major calyx bifurcates or trifurcates once more into **minor calyces**. A longitudinal section

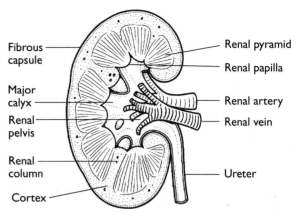

**Figure 7.4**  The cortex of the kidney contains the glomeruli and the convoluted tubules which extend down through the medulla as renal columns. The medulla contains renal pyramids which contain collecting tubules. These point at renal papillae into the minor calyces.

through the kidney reveals there is an outer **cortex** and an inner **medulla**. The medulla consists of between five and ten **renal pyramids** whose apices (or **papillae**) point towards the minor calyces.

# The suprarenal glands

Both right and left kidneys are closely related to the suprarenal glands at their upper poles. The right suprarenal gland squeezes medially underneath the inferior vena cava (Fig. 7.5) and part of it lies behind the bare area of the liver. The left suprarenal gland extends down towards the hilum of the kidney more than the gland on the right. Both right and left suprarenal glands are enclosed by the posterior and anterior layers of renal fascia but they have their own

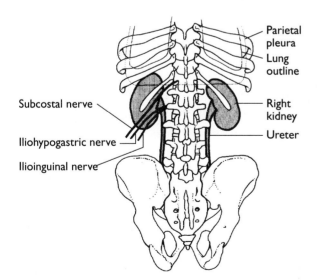

**Figure 7.3**  Posteriorly the kidneys relate to the 12th rib, the parietal pleura through the diaphragm and costodiaphragmatic recess, the subcostal nerve, iliohypogastric and ilioinguinal nerves. The ureters run down to the pelvis in line with the tips of the lumbar transverse processes.

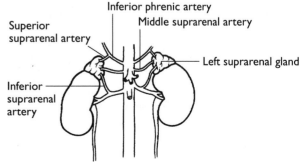

**Figure 7.5**  The suprarenal glands receive arterial blood from three sources.

compartments within this fascia that separate them from the renal capsule and kidney. Each suprarenal gland receives three arteries (Fig. 7.5). The **superior suprarenal artery** is a branch of the inferior phrenic artery. The **middle suprarenal artery** is a direct branch of the abdominal aorta. The **inferior suprarenal artery** is a branch of the renal artery. Just one suprarenal vein drains into the inferior vena cava on the right but into the renal vein on the left.

# The peritoneal cavity and the mesenteries

At first sight, the layout of the viscera in the abdominal cavity looks very complicated. This is because a fundamentally simple plan has undergone some obscure changes during development. It is worth following some of these changes through, in a very simple way, so that we can better understand what we see in the adult abdominal cavity. First of all we will look at the development of the foregut and its derivatives and their rotations. Then we will consider the development and rotation of the midgut and hindgut.

## Development of the foregut

Just as the pleural or pericardial cavities are lined with a membrane whose inner surface consists of serous mesothelium, so the peritoneum forms a **peritoneal cavity** within the abdomen that is lined by serous endothelium (Fig. 7.6). Initially, a long thin gut tube lies in front of the aorta behind the peritoneal sac. The part of the peritoneum that is in contact with the gut tube is called the **visceral peritoneum**. The rest of the peritoneum that lines the walls of the abdominal cavity is called the **parietal peritoneum**. Soon, the gut tube moves forwards into the abdominal cavity drawing with it a double-layered fold of peritoneum that remains anchored at the back in front of the aorta. This double-layered 'stretched out' portion of peritoneum is called the **dorsal mesentery**. The **root** of the dorsal mesentery is the part where the mesentery becomes continuous with the parietal peritoneum at the posterior abdominal wall (Fig. 7.6). A **mesentery** or **ligament** in the abdominal cavity is then simply a double-layered fold of peritoneum.

Two derivatives of the foregut begin to develop in the dorsal mesentery high up in the abdominal cavity behind the future stomach (Fig. 7.7(a)). These are the dorsal part of the pancreas and the spleen. Two other gut derivatives also begin to push out in front of the future stomach within the **ventral mesentery**. These are the ventral part of the pancreas and the liver. They extend within the ventral mesentery to the anterior abdominal wall. Unlike the dorsal mesentery, the ventral mesentery exists only in the upper part of the abdominal cavity where the liver and ventral bud of the pancreas are developing. There is, therefore, a free lower border to the ventral mesentery and free passage between the left and right sides of the abdominal cavity below it (Fig. 7.7(a)).

The structures developing within the ventral

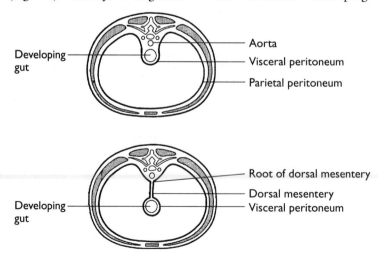

**Figure 7.6** The developing gut tube at first lies behind the peritoneal cavity but then extends forwards drawing with it a covering of visceral peritoneum. Only the dorsal mesentery then connects it to the posterior abdominal wall. (After Stern JT. *Essentials of Gross Anatomy*. Philadelphia: FA Davis, 1988.)

(a)

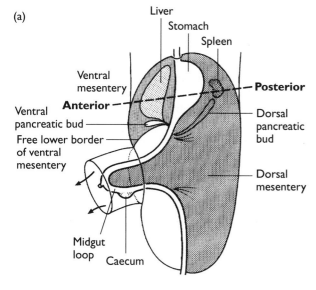

(b)

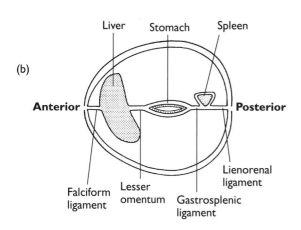

***Figure 7.7*** In (a) the primitive gut tube has differentiated into a recognizable stomach and midgut loop with a caecum. A ventral mesentery with a free lower border contains the developing liver and ventral bud of the pancreas. The dorsal mesentery contains the developing spleen and dorsal pancreatic bud. In (b) a section through the ventral and dorsal mesenteries high up (in the plane of the dotted line) shows the named ligaments that connect the liver, stomach and spleen to each other and to the abdominal walls. (After Green JH and Silver PHS. *An Introduction to Human Anatomy.* New York: Oxford Medical Publications, 1981.)

together and connect their peripheries to the anterior or posterior abdominal wall or the diaphragm above. The **falciform ligament** with its free lower border connects the liver to the anterior abdominal wall and diaphragm. The **lesser omentum** connects the liver to the so-called **lesser curvature** of the stomach. The spleen divides the dorsal mesentery of the stomach (which is attached to its so-called **greater curvature**) into a part between it and the stomach called the **gastrosplenic ligament** and a part between it and the left kidney called the **lienorenal ligament**. The dorsal mesentery that passes from the greater curvature of the stomach *below* the gastrosplenic ligament to the root of the mesentery on the posterior abdominal wall is known as the **greater omentum**.

The stomach and its attached mesenteries containing the various gut derivatives then rotate to the right side of the abdominal cavity (Fig. 7.8). If you hold your left hand in front of your abdomen (fingers pointing forwards) so that your thumb sticks up (to represent the oesophagus) then the web of skin between the thumb and the first finger represents the lesser curvature of the stomach (Green JH and Silver PHS (1981) *An Introduction to Human Anatomy.* New York: Oxford Medical Publications). The border formed by the tips of the fingers and the side of the little finger and palm then represents the greater curvature of the stomach. Turn your hand to the right so that the palm now faces your abdomen and you have mimicked what happens in the embryo. (You can't do it with your right hand because your wrist won't bend the other way.) This rotation of the stomach and new position of the greater curvature on the left, and lesser curvature on the right affects the position of the spleen which is carried further out to the left as is the dorsal mesentery whose root now shifts to the left of the aorta (Fig. 7.8).

The pylorus of the stomach and the future duodenum are, on the other hand, carried over to the right. The dorsal mesentery of the future duodenum rotates round to the right and comes to lie flat against the parietal peritoneum on the posterior wall of the abdomen. The parietal peritoneum and (now posterior) layer of the dorsal mesentery that are in contact then 'melt away' and the duodenum becomes part of the posterior abdominal wall. The single remaining layer of the mesentery covering the front of the duodenum then behaves exactly like parietal peritoneum so that the duodenum becomes a **secondarily retroperitoneal** structure. The pancreas that was

mesntery come to divide the region in front of the future stomach into zones. Likewise, the structures in the dorsal mesentery behind the future stomach divide it into distinct zones (Fig. 7.7(a) and (b)). The mesentery of the zones connect each of the structures

(a)

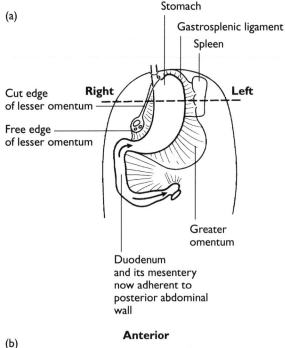

(b)

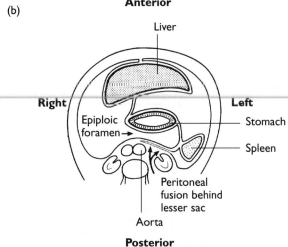

**Figure 7.8**   The foregut is seen from the front in (a). The stomach has rotated to the right along with the duodenum. The mesentery of the duodenum has rotated against and fused to the parietal peritoneum on the posterior abdominal wall. The greater omentum still ties the stomach tightly to the posterior abdominal wall. The lesser omentum is cut at its connection with the liver. The lower free edge is still preserved. A transverse section (b) shows how the spleen has swung out to the left and the stomach rotated to enclose the lesser omentum behind it. The epiploic foramen is the entrance into the lesser sac. (After Green JH and Silver PHS. *An Introduction to Human Anatomy.* New York: Oxford Medical Publications, 1981.)

within the dorsal mesentery of the stomach and duodenum also becomes retroperitoneal in this way at the same time.

You will recall that the part of the mesentery running from the greater curvature of the stomach to the posterior abdominal wall is called the greater omentum. If the greater omentum remained short in its new rotated position it would become tight when the stomach was filled with food. So its middle portion increases in length and hangs down as far as the pelvis in preparation for anticipated distension of the stomach while feeding (Fig. 7.9).

As a consequence of the rotation of the stomach to the right the 'old right' surface of the stomach now becomes its posterior surface. The space between the posterior abdominal wall and the posterior surface of the stomach is called the **omental bursa** or **lesser sac**. The liver divides the ventral mesentery of the stomach into two portions. One runs between the liver and the diaphragm and anterior abdominal wall as far down as the umbilicus. This is called the falciform ligament. The other runs between the lesser curavture of the stomach and the liver itself. This is called the lesser omentum. Both the falciform ligament and the lesser omentum retain the free lower border of the ventral mesentery. The free edge of the lesser omentum has turned over to the right and there is an opening into the lesser sac called the **epiploic foramen** behind it. Since the duodenum is now retroperitoneal, the free edge of the lesser omentum is tied to the posterior abominal wall at its lower end. The ligaments that

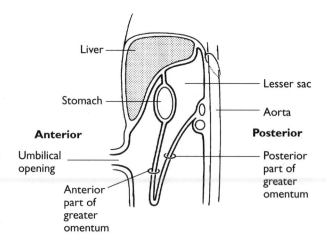

**Figure 7.9**   Elongation of the anterior and posterior parts of the greater omentum allows the stomach to move and distend more freely. (After Stern JT. *Essentials of Gross Anatomy.* Philadelphia: FA Davis, 1988.)

now attach the liver to the diaphragm and posterior abdominal wall limit the boundaries of the lesser sac above and posteriorly.

# Development of the midgut and hindgut

The midgut consists of a part of the **duodenum**, the **jejunum**, the **ileum** and rather more than half of the large intestine or **colon**. It has its own blood supply in the form of the superior mesenteric artery. The hindgut consists of the distal third of the transverse colon and the sigmoid colon, rectum and upper anal

canal. The hindgut too has its own artery, the inferior mesenteric artery.

During development a 'midgut loop' first projects out of the abdominal cavity and into an extension in the umbilical opening, drawing its mesentery and blood vessel with it (Fig. 7.10(a)). A **caecum** appears at the junction of the ileum and colon so that from this time on it is possible to follow the development of the small and large intestines separately. The axis of the midgut loop corresponds with that of the superior mesenteric artery. The caecum and future large intestine then rotate upwards and in front of the future small intestine (Fig. 7.10(b)). This rotation, you will note, is anticlockwise around the axis. Remember that the transverse colon ends up in front

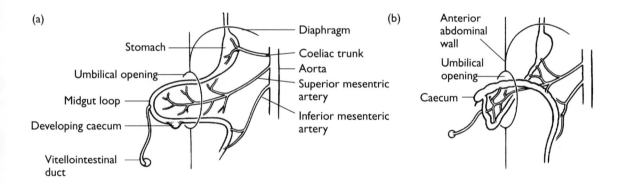

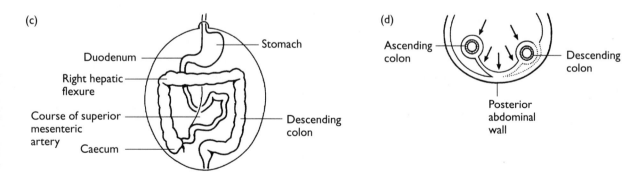

**Figure 7.10**   (a) The midgut loop and superior mesenteric artery extend through the umbilical opening. The caecum and colon swing up in front of the duodenum and small bowel (b). The descending colon is then pushed to the left; next the small bowel crowds in on top of it and finally the ascending colon returns to the abdominal cavity (c). The roots of the mesenteries of the ascending and descending colon are flattened against the parietal peritoneum on the posterior abdominal wall, to which they fuse and become secondarily retroperitoneal (d). (After Green JH and Silver PHS. *An Introduction to Human Anatomy.* New York: Oxford Medical Publications, 1981.)

(a)

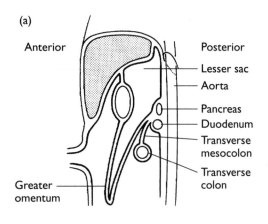

Anterior

Posterior

Lesser sac

Aorta

Pancreas

Duodenum

Transverse mesocolon

Transverse colon

Greater omentum

(b)

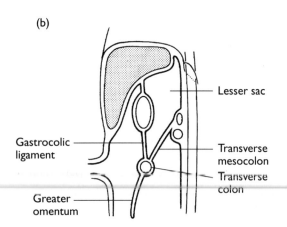

Lesser sac

Gastrocolic ligament

Transverse mesocolon

Transverse colon

Greater omentum

**Figure 7.11**   Initially both the transverse mesocolon and the posterior part of the greater omentum have their separate roots over the pancreas and duodenum. These fuse to form the transverse mesocolon. (After Stern JT *Essentials of Gross Anatomy.* Philadelphia: FA Davis, 1988.)

of the duodenum and small intestine and you will not get the direction of the rotation mixed up.

The gut then returns to the abdomen. The first part to return to the posterior wall of the abdominal cavity is the **descending colon**. It gets pushed over to the left by the small intestine, which is crowding back on top of it into the central region of the cavity. The root of the mesentery of the small intestine comes to lie obliquely across the posterior wall of the abdominal cavity, from the junction of the duodenum with the jejunum to the terminal part of the ileum. The last structures to return are the caecum and appendix. At this time the caecum and appendix lie close to the liver and gall bladder. Rarely, this relationship persists into adulthood, but normally the caecum descends, drawing out the **ascending colon** behind it until it reaches the right iliac fossa. The mesenteries of the ascending and descending colon come to lie flat on the parietal peritoneum on the posterior abdominal wall. The colon is pushed against the wall in this way by the mass of small intestine crowding around it. The mesenteries of the ascending and descending colon still contain the blood vessels, lymphatics and nerves supplying these parts of the gut. Slowly, the adherent layers of peritoneum and mesenteries are absorbed and disappear completely, obliterating a part of the peritoneal cavity behind them. Thus, the ascending and descending colon become retroperitoneal and immobile in exactly the same way as the duodenum and pancreas. A part of the hindgut between the descending colon and the rectum keeps its dorsal mesentery. This is the **sigmoid colon**.

The root of the mesentery of the transverse colon lies in front of the pancreas. The mesentery itself does not adhere to the posterior abdominal wall but fuses instead with that part of the dorsal mesentery of the stomach which makes up the greater omentum (Fig. 7.11). The roots of these mesenteries in any case by now both lie very close together over the pancreas and duodenum, and their fusion is predictable. When fused, the mesentery passing from the transverse colon to the posterior abdominal wall is known as the **transverse mesocolon**. Mobilization of retroperitoneal portions of the gut during surgery requires that the regions of secondary adhesion are cut and freed leaving the 'original' mesentery and its vessels and nerves intact.

chapter

8

# The Liver, Gall Bladder and Gastrointestinal Tract

## The liver and gall bladder

The liver lies beneath the dome of the diaphragm on the right as well as under the central tendinous portion of the diaphragm beneath the inferior surface of the heart. The part of the liver adjacent to the diaphragm is known as its diaphragmatic surface. A number of abdominal structures are in contact with the liver against its visceral surface. The wedge shape of the liver is determined largely by the contours of the diaphragm and those structures that lie against it.

Look now at the diagrams of the diaphragmatic and visceral surfaces of the liver (Fig. 8.1). The falciform ligament and a fissure (formed by the **ligamentum venosum** that runs across the visceral surface of the liver) appear to divide the liver into right and left lobes. However, the right and left hepatic arteries, right and left hepatic ducts and branches of the portal vein, that enter the liver at the **porta hepatis** on the visceral aspect, do not divide to pass either to the left or right of these ligaments. It seems that a better division of the liver into **functional** right and left **lobes** can be defined by a line drawn between the fossa for the gall bladder in front to the fossa with the inferior vena cava in it behind.

The ligamentum venosum is a remnant of the **ductus venosus**. This duct shunts oxygenated blood from the umbilical vein past the liver and into the inferior vena cava before birth. The **ligamentum teres** is the obliterated remnant of the left umbilical vein. The lesser omentum merges with the visceral peritoneum of the liver in the floor of the fissure formed by the ligamentum teres and ligamentum venosum. The

part of the liver between the fissure for the ligamentum venosum and the fossa for the inferior vena cava is called the **caudate lobe** of the liver. The falciform ligament merges with an extension of the fissure for the ligamentum teres at the anterior edge of the liver. The part of the liver between the fossa for the gall bladder and this anterior part of the fissure is called the **quadrate lobe** of the liver.

The visceral peritoneum of the liver that formed from the two layers of the ventral mesentery does not completely encapsulate the liver. The falciform ligament that represents the double-layered ventral mesentery can be followed on to the anterior surface of the liver from its attachments to the anterior abdominal wall and diaphragm (Fig. 8.1). Superiorly, the falciform ligament splits into right and left halves as it meets the liver. The upper edges run around the high dome of the liver like a collar, reflecting on to the diaphragm as they go, and meeting up at the back again. On the right side of the liver the **coronary ligament** encloses a **bare area** of the liver. The bare area of the liver is in direct contact with the diaphragm posterosuperiorly and also with the inferior vena cava and a part of the right suprarenal gland posteriorly. At the right edge of the coronary ligament, the anterior and posterior margins of visceral peritoneum make contact and run together as a double layer in a straight line for a while. This is known as the right **triangular ligament**. The left **triangular ligament** runs as a double layer to the left of the falciform ligament.

Two **subphrenic pockets** or **spaces** lie between the diaphragm and the liver on either side of the falciform ligament (Fig. 8.2). These are important pockets where pus can collect in a patient with

(a)  Anterior view

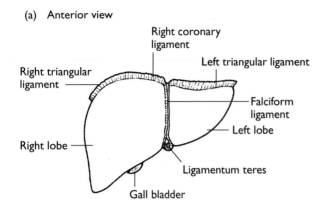

(b)  Inferior view

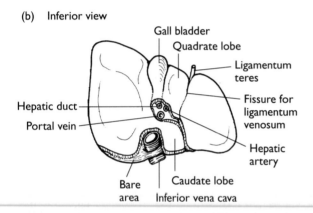

(c)  Superior view

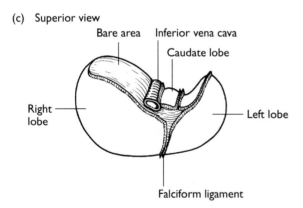

**Figure 8.1**   The falciform ligament runs between the anterior abdominal wall and the liver. The ligamentum teres, in its lower border, continues as the ligamentum venosum in the groove on the undersurface of the liver. The quadrate lobe lies between this groove and the gall bladder. The caudate lobe lies between it and the inferior vena cava. Note the bare area that lies against the diaphragm.

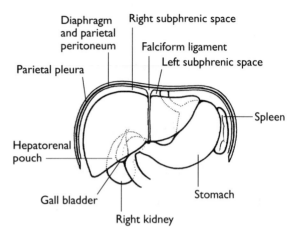

**Figure 8.2**   The hepatorenal pouch lies between the right kidney and the posterior surface of the liver. The right and left subphrenic spaces lie beneath the diaphragm on either side of the falciform ligament.

intra-abdominal sepsis. The left and right subphrenic spaces can be explored by passing a hand up and over the anterior surface of the liver on either side of the falciform ligament, as far as it will go, before meeting the coronary or triangular ligaments. The liver overlies the right kidney posteriorly. A **subhepatic** or **hepatorenal space** or **pouch** may also collect and harbour pus. This space lies between the inferior visceral surface of the liver and the right kidney, where it has important relations to the duodenum and hepatic flexure of the colon. Pus can collect here in the supine patient. The liver is also in contact with the oesophagus, stomach on the left, and the duodenum and right hepatic flexure of the colon as well as the right kidney and right suprarenal gland.

## The gall bladder

The gall bladder (Fig. 8.3) lies directly against the liver in the **cystic fossa**. The gall bladder concentrates bile some ten to 12 times by absorbing water from it. The gall bladder is covered on its external surface by visceral peritoneum that also runs over the liver. The gall bladder has a **fundus** that lies at the margin of the liver and which is opposite the tip of the 9th costal cartilage. The **body** of the gall bladder narrows to a **neck** which is continuous with the **cystic duct**. The body of the gall bladder lies against the transverse colon and in front of the first part of the duodenum. The cystic duct from the gall bladder runs to join the **common hepatic duct**, a duct formed close to the

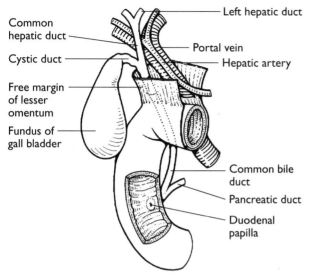

**Figure 8.3** The common bile duct, hepatic artery and hepatic portal vein lie in the free margin of the lesser omentum. The duodenal papilla opens into the second part of the duodenum. (After Hall-Craggs ECB. *Anatomy as a Basis for Clinical Medicine*. Munich: Urban and Schwarzenberg, 1990.)

porta hepatis by fusion of the right and left hepatic ducts. After the cystic duct and the common hepatic duct have joined up they are known as the **common bile duct**. The cystic duct, common bile duct, portal vein and hepatic artery are all found between the leaves of the lesser omentum in the free margin of the lesser omentum. The common bile duct passes behind the first part of the duodenum and into the substance of the head of the pancreas. Here it joins with the pancreatic duct at the **hepatopancreatic ampulla** and opens into the duodenum at the **duodenal papilla** in the wall of the second part of the duodenum.

The blood supply to the liver and gall bladder comes from the coeliac trunk (Fig. 8.4). The coeliac trunk sprouts three main arteries close to its origin, the **common hepatic artery**, the **splenic artery** and the **left gastric artery**. The common hepatic artery gives off the right gastric artery and the gastroduodenal artery before finally dividing into right and left hepatic arteries close to the porta hepatis. The right hepatic artery usually passes behind the common hepatic duct as it nears the porta hepatis and here it usually gives off the **cystic artery** to supply the gall bladder. There is nevertheless much variation in the arterial supply to the liver and gall bladder, and also in the arrangement of the cystic and hepatic ducts, and each individual

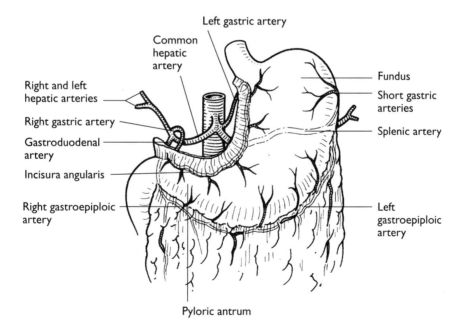

**Figure 8.4** The coeliac trunk arises in the midline of the aorta. It gives off the splenic artery, the common hepatic artery and the left gastric artery. The right gastric and gastroduodenal arteries arise from the common hepatic artery. (After Hall-Craggs ECB. *Anatomy as a Basis for Clinical Medicine*. Munich: Urban and Schwarzenberg, 1990.)

case must be carefully explored in its own right and not just presumed to follow the common pattern.

# The stomach, duodenum, pancreas and spleen

## The stomach

The oesophagus passes through the diaphragm at the level of T10. We are aware now that the right crus of the diaphragm curls around the oesophageal opening and can assist the **cardiac sphincter** of the stomach in preventing regurgitation of food back up the oesophagus. The **fundus** of the stomach (Fig. 8.4) rises up under the dome of the diaphragm on the left and is often seen to contain swallowed air on radiographs taken in the standing position. The **body** of the stomach forms most of its mass and is continuous with the **antrum** and **pyloric region** to the right of the midline. The junction of the body with the antrum is marked by a groove or notch in the lesser curvature called the **incisura angularis**. Inside, the mucosa of the stomach is thrown into longitudinal folds, or **rugae**, that follow its contours towards the duodenum. The **pyloric sphincter**, unlike the cardiac sphincter, is thickened by circular smooth muscle which controls the flow of the stomach contents into the duodenum. The stomach is fixed at the cardiac sphincter (the surface marking of this point is just to the left of the 6th costal cartilage in the midline). The stomach is also fixed at the pyloric sphincter (the surface marking of the pyloric sphincter is just to the right of the midline in the transpyloric plane). Between these points the stomach is free to writhe around and expand as far as the greater and lesser omenta will allow.

Anteriorly, the stomach relates to the liver, diaphragm and anterior abdominal wall. Posteriorly it lies on a bed consisting of the left crus of the diaphragm, the pancreas, the left kidney and suprarenal gland as well as the transverse colon and the spleen.

## Blood supply and lymphatic drainage of the stomach

Blood supply to the stomach comes from the coeliac trunk (Fig. 8.4). The **right** and **left gastric arteries** run along the lesser curvature of the stomach in the lesser omentum. The right gastric artery is one of the branches of the common hepatic artery. The left gastric artery is a direct branch of the coeliac trunk. The

**right** and **left gastroepiploic arteries** run along the greater curvature of the stomach in the greater omentum. The right gastroepiploic artery comes from the other major branch of the common hepatic artery, the **gastroduodenal artery**, and arises from it behind the first part of the duodenum. The left gastroepiploic artery arises from the other artery that runs behind the stomach, the **splenic artery**. The splenic artery is a direct branch of the coeliac trunk and as it passes behind the greater curvature of the stomach it gives off the left gastroepiploic artery. This runs down the greater curvature to meet its right namesake the right gastroepiploic artery. The splenic artery also gives rise to the **short gastric arteries** that run up the greater curvature of the stomach to the fundus.

Gastric veins all drain to the portal vein, and therefore back to the liver. It is worth emphasizing here that venous blood from the distal portion of the oesophagus also drains back to the portal vein via lower oesophageal veins. Any venous obstruction in the portal vein or liver may result in **oesophageal varices** and potential bleeding into the lumen of the lower oesophagus.

Lymph from the stomach drains in one of three ways: upwards towards the cardiac sphincter, downwards to the pyloric sphincter, or left and laterally towards the spleen. Chains of lymph nodes lie along the arteries to the stomach and are named after them: coeliac trunk group, hepatic group, splenic group, left gastric group etc. All of these groups eventually drain into the thoracic duct via the coeliac group of nodes. Lymphatic drainage of the gut, generally, follows the arterial tree back towards the aorta and the para-aortic lymph nodes. From here on it makes its way to the **cysterna chyli** and the thoracic duct.

## Autonomic nerve supply to the stomach

Both sympathetic nerves from the thoracic splanchnic nerves (T5 to T9) and parasympathetic nerves from the vagal trunks innervate the stomach. The right vagus contributes largely to the plexus on the posterior surface of the lower third of the oesophagus. The left vagus contributes largely to the plexus on the anterior surface of the oesophagus in its lower third. The parasympathetic oesophageal plexuses reform as the anterior and posterior gastric nerves as they enter the abdomen, and these (especially the anterior) pass to the fundus, body and pylorus of the stomach. Sympathetic fibres synapse in the **coeliac ganglia** around the coeliac trunk and then stream on to the stomach with the blood vessels. Sympathetic nerves inhibit peristalsis and close the pyloric sphincter. Parasympathetic nerves increase peristalsis, relax the pylorus and are secretor motor to the secretory glands of the stomach mucosa.

## The duodenum

The **first part** of the duodenum (Fig. 8.5) lies at the level of L1 and is continuous with the stomach at the pylorus. The first part of the duodenum runs backwards. The lesser omentum is attached to its upper surface here. The lesser omentum has a free border that runs from the first part of the duodenum to the liver. The rest of the duodenum, as we have seen, is retroperitoneal and is therefore immobile. The duodenum lies over to the right like a horse-shoe with the head of the pancreas tucked into the curvature. The **second part** of the duodenum descends and crosses over the hilum of the right kidney. It has the transverse mesocolon in front of it. It is into the second or descending part of the duodenum that the combined pancreatic and biliary duct enters the lumen at the duodenal papilla. The **third part** of the duodenum begins as it levels out into the lower horizontal part of the curve. This part of the duodenum lies just above the level of the umbilicus at the level of L3. The third part of the duodenum has both the inferior vena cava and aorta behind it as well as the right gonadal vein and artery. Behind all of these structures is the right ureter lying on the right psoas muscle. Anteriorly, the superior mesenteric vessels

pass in front of the third part of the duodenum. The **fourth part** of the duodenum ascends again and rises out from behind the peritoneum to join with the jejunum at the duodenojejunal junction. It is actually suspended here from the vertebral bodies close to the origin of the right crus by a ligament. This ligament may have some muscle fibres in it and is called the **suspensory ligament of the duodenum** or **ligament of Treitz**.

There are folds of peritoneum around the fourth part of the duodenum that form pouches as it re-enters the peritoneal cavity. These are the **superior** and **inferior duodenal fossae**. Very occasionally, small bowel may get trapped in one of these fossae, forming an internal hernia.

### Blood supply to the duodenum

The duodenum gets part of its blood supply from the artery to the foregut and part from the artery to the midgut. The gastroduodenal branch of the common hepatic artery divides into the right gastroepiploic artery, that we have already studied, and the **superior pancreaticoduodenal artery**. Each of these arteries is derived from the coeliac trunk. The first branch of the superior mesenteric artery is the **inferior pancreaticoduodenal artery**. As we have seen, the superior mesenteric artery and vein pass in front of the horizontal third part of the duodenum and it is here that the inferior pancreaticoduodenal artery is given off.

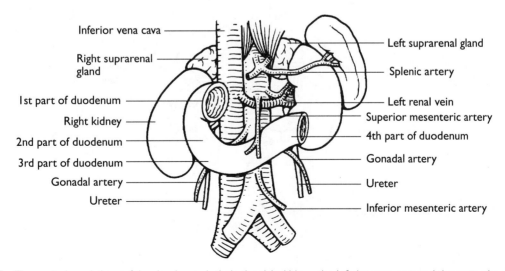

**Figure 8.5**   The posterior relations of the duodenum include the right kidney, the inferior vena cava and the aorta, the right ureter and the right gonadal vein. The superior mesenteric vessels run in front of the third part of the duodenum.

## The pancreas

The pancreas consists of a **head**, a **neck**, a **body** and a **tail** (Fig. 8.6). The pancreas is triangular in cross-section with its base applied to the posterior abdominal wall. Apart from its tail it is retroperitoneal. The head and neck of the pancreas lie within the curvature of the duodenum on the right and have important structures passing behind them. These include the aorta and inferior vena cava, the common bile duct and portal vein. The superior mesenteric artery also squeezes out from behind the pancreas to run on to the front of the third part of the duodenum. In fact there is a small part of the pancreas (the bit that developed in the ventral mesentery) that hooks behind the superior mesenteric artery is called the **uncinate process** of the pancreas. But, in reality, the superior mesenteric artery runs from its origin on the front of the aorta behind the pancreas and not through its substance. The tail of the pancreas lies between the leaves of the lienorenal ligament, which continues on to the hilum of the spleen. In front of the pancreas the **transverse mesocolon** rises from the parietal peritoneum in the region of the body and head.

## Blood supply and ducts of the pancreas

Blood to the pancreas comes from the superior and inferior pancreaticoduodenal arteries and also from branches of the **splenic artery** as it runs along the upper border of the pancreas towards the spleen. The duct of the pancreas runs through the gland to join with the common bile duct in the head of the pancreas. The parts of the pancreas that developed in the ventral and dorsal mesenteries each have their own duct. Usually these join together before meeting up with the common bile duct, but sometimes the **accessory duct** from the uncinate process enters the duodenum on its own above the level of the main duct.

## The spleen

The spleen lies posterolaterally on the diaphragm in the long axis of the 10th rib on the left side of the abdominal cavity (Figs 8.5 and 8.6). It has a smooth diaphragmatic surface. Its anterior border is notched and when enlarged (but only when enlarged) it can be felt at the edges of the 9th, 10th and 11th ribs on deep inspiration. The visceral surface of the spleen is divided into three regions by a 'Y-shaped' central region (Fig. 8.7). The stomach lies against the spleen

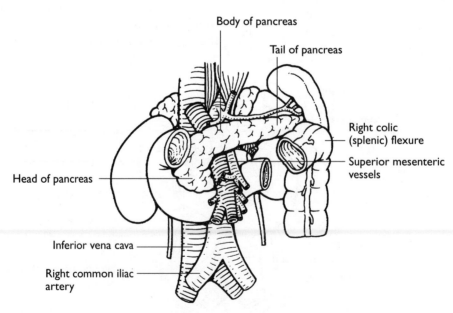

**Figure 8.6**   The head of the pancreas lies in the 'C'-shaped curve of the duodenum. The body of the pancreas lies across the posterior abdominal wall and the tail within the lienorenal ligament and in contact with the spleen.

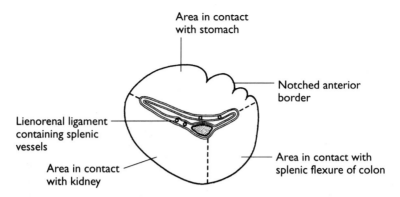

**Figure 8.7** The visceral aspect of the spleen is in contact with the stomach, the splenic flexure of the colon and the left kidney. It lies against the 9th, 10th and 11th ribs on the left along the axis of the 10th rib.

above in one region, the kidney lies against it posteriorly and the splenic flexure of the colon lies against the anterior part of the visceral surface. The spleen is connected to the parietal peritoneum over the renal fascia by the lienorenal ligament in front of the left kidney. The tail of the pancreas and the splenic artery and vein run within the lienorenal ligament to and from the hilum of the spleen.

The splenic artery is a direct branch of the coeliac trunk. It is a large tortuous vessel that runs along the upper border of the pancreas and then within the lienorenal ligament to the spleen. The splenic vein is also large and is one of the main tributaries of the portal vein. Note, therefore, that blood from the spleen drains back to the portal system of the liver.

There is also a peritoneal attachment between the stomach and the spleen, the gastrosplenic ligament. The gastrosplenic ligament contains the short gastric arteries that run from the splenic artery to the fundus of the stomach. The relation of these ligaments to the organs in contact with the spleen can be seen in Figures 8.6 and 8.7, and are responsible for creating the Y-shaped pattern on the central portion of the visceral surface of the spleen.

## The small intestine

The jejunum and the ileum together make up the small intestine. There is, however, no absolute demarcation between them. They are both suspended in the abdominal cavity from the posterior abdominal wall by a mesentery. The root of the mesentery runs obliquely across the posterior abdominal wall from

the duodenojejunal junction to the **ileocaecal junction**. The root of the mesentery is little more than 15 cm long so that the great 5 m length of small bowel that it suspends is thrown into numerous coils and folds. The superior mesenteric artery runs into the root of the mesentery high up and gives off jejunal and ileal branches to the left and branches to the large bowel on its right side. The jejunal and ileal branches run parallel with each other into the mesentery. Where the mesentery is shortest, near the upper jejunum, just one set of vascular arches develops between adjacent vessels (Fig. 8.8). Where the mesentery is longer two or even three or four generations of vascular arches develop in the mesentery. The terminal branches of all the arches are straight and run into the wall of the jejunum or ileum parallel with each other.

The jejunum looks more vascular than the ileum. It has a thicker wall and its lumen is thrown into circular folds called **plicae circulares**. There is usually fat in the mesentery of both ileum and jejunum but this does not run up as close to the jejunum as it does to the ileum. The ileum is less vascular, feels thinner between the finger and thumb, and has many more lymphoid follicles (**Peyer's patches**) in its wall than the jejunum.

You will remember that the axis of the midgut loop corresponds with the axis of the superior mesenteric artery. Directly opposite this, the **vitellointestinal duct** extended from the small intestine of the embryo to the yolk sac. Occasionally in adults there is an abnormal persistence of this duct in the form of a pouch or diverticulum called **Meckel's diverticulum** (Fig. 8.9). It occurs about 0.5 m from the ileocaecal junction

## (a) Proximal jejunum

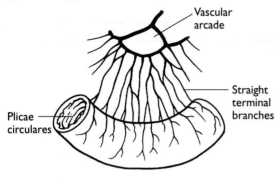

Vascular arcade

Straight terminal branches

Plicae circulares

## (b) Distal ileum

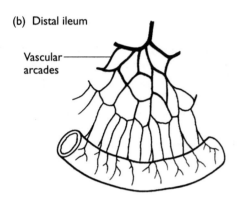

Vascular arcades

**Figure 8.8** The jejunum (a) is thicker, more vascular and has fewer vascular arcades than the ileum (b). The inner surface of the jejunum is thrown into folds called plicae circulares. The inner surface of the ileum has many Peyer's patches.

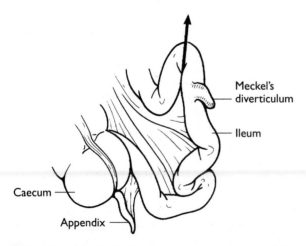

Meckel's diverticulum

Ileum

Caecum

Appendix

**Figure 8.9** Meckel's diverticulum, when present, lies in the the axis of the superior mesenteric artery. It is an outpouching of the ileum that persists at the original position of the vitellointestinal duct of the embryo.

and can become inflamed and indistinguishable in its clinical presentation from an inflamed appendix.

# The large intestine

The outer longitudinal muscular coat of the bowel is reduced to three flat muscular bands only in the large intestine (Fig. 8.10). These are called **taeniae coli**. The taeniae coli run across the more voluminous circular bands of muscle and throw them into pouches or sacs called **haustra**. Small sacs of fat, **appendices epiploicae**, also hang from the large intestine and together with the taeniae coli easily distinguish small bowel from large bowel.

The caecum (Fig. 8.11) is a large dilated sac at the proximal end of the large intestine which lies in the right iliac fossa on the iliacus and psoas muscles. The caecum is retroperitoneal and has a retrocaecal peritoneal pocket behind it. At birth the appendix is attached to the apex of the caecum but during growth the base of the appendix is pushed to lie in a posteromedial position on the caecum. The taeniae coli always converge on to the base of the appendix and this is a useful way of finding it. The ileum also opens into the caecum just above the opening for the appendix (Fig. 8.11). There are rounded lips or flaps which form the **ileocaecal valve** but they appear to have little impact on the flow of bowel contents either way. The base of the appendix has an important surface marking which is said to correspond to the 'junction of the lateral and middle third of a line joining the umbilicus and the right anterior superior iliac spine'. This is **McBurney's point**. The tip of the appendix, however, may lie in one of many places and may be **retrocaecal**, **paracaecal**, **subcaecal**, **retrocolic** or even hang into the true pelvis.

The blood supply to the appendix comes from the **ileocolic** branch of the superior mesenteric artery (Fig. 8.11). Its caecal branch gives off anterior and posterior caecal vessels. The **appendicular artery**, a branch of the posterior caecal artery, runs *under* the terminal portion of the ileum and then in the free edge of the **mesoappendix**.

The ascending colon is retroperitoneal and immobile, and extends from the right iliac fossa to the right colic or hepatic flexure. The transverse colon hangs from the transverse mesocolon usually as far as the umbilicus but may descend further even to the pelvis.

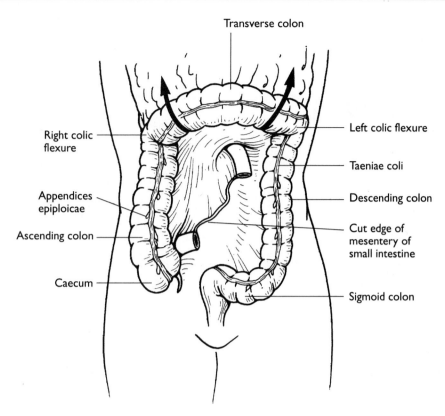

**Figure 8.10** The colon has three longitudinal bands of muscle called taeniae coli. The ascending and descending colon are secondarily retroperitoneal but the transverse colon is mobile and attached to the posterior abdominal wall by the transverse mesocolon. (After Hall-Craggs ECB. *Anatomy as a Basis for Clinical Medicine.* Munich: Urban and Schwarzenberg, 1990.)

The transverse colon extends to the left colic or splenic flexure which is at a higher level than the right colic flexure. The descending colon is once again retroperitoneal and is tethered laterally to the diaphragm by the **phrenicocolic fold** at its origin. It also crosses the lower pole of the left kidney and relates to the spleen. The sigmoid colon begins at the inlet of the true pelvis and has a surprisingly long mesentery. One limb of its mesentery runs along the brim of the pelvis and the other runs down to the midline.

Lateral to the ascending and descending colon there are two **paracolic gutters**, each of which runs up between the lateral abdominal wall and the outer margin of the colon. Above, the right paracolic gutter communicates with the hepatorenal pouch but the left paracolic gutter ends at the phrenicocolic fold or ligament.

The blood supply to the colon comes for the most part from the **right colic** and **middle colic** branches of the **superior mesenteric artery** which arise on the right side of this artery (Fig. 8.12). These vessels sup-

ply the colon as far as the distal third of the transverse colon. The right colic artery approaches the ascending colon behind the peritoneum running over the right gonadal vessels and the right ureter. The middle colic artery runs into the transverse mesocolon close to the lower border of the pancreas. The last part of the transverse colon and the descending colon get their blood supply from the **inferior mesenteric artery**. This artery is smaller than the superior mesenteric artery and emerges beneath the third part of the duodenum (Fig. 8.13). It gives off a **left colic artery** that rises up behind the peritoneum (crossing the left ureter and left gonadal vessels) to supply the region of the splenic flexure. There is a long anastomotic vessel that runs in the transverse mesocolon along the edge of the colon and meets up with branches from the left colic artery. This is called the **marginal artery (of Drummond)**. Two or three **sigmoid arteries** run down to the descending colon and the sigmoid colon. The inferior mesenteric artery then continues into the pelvis as the **superior rectal artery**.

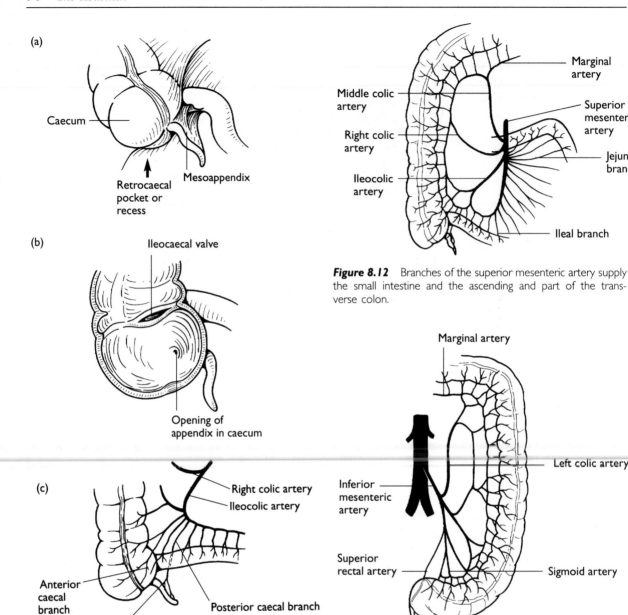

(a)

Caecum

Retrocaecal pocket or recess

Mesoappendix

(b)

Ileocaecal valve

Opening of appendix in caecum

(c)

Right colic artery

Ileocolic artery

Anterior caecal branch

Posterior caecal branch

Appendicular artery

**Figure 8.11**  All three taeniae coli converge at the base of the appendix (a). Within the caecum (b) the ileocaecal valve and the base of the appendix open into the lumen of the caecum. The appendicular artery is a branch of the posterior caecal branch of the right colic artery (c).

Marginal artery

Middle colic artery

Right colic artery

Ileocolic artery

Superior mesenteric artery

Jejunal branch

Ileal branch

**Figure 8.12**  Branches of the superior mesenteric artery supply the small intestine and the ascending and part of the transverse colon.

Marginal artery

Inferior mesenteric artery

Left colic artery

Superior rectal artery

Sigmoid artery

**Figure 8.13**  Branches of the inferior mesenteric artery supply the distal part of the transverse colon, the descending and sigmoid colon, and the upper part of the rectum.

## The rectum and anal canal

It makes sense here to give an account of the rectum and anal canal, even though some details are best considered together with an account of the pelvic floor. The **rectum** follows the sacral flexure and ends at the anal canal where it turns abruptly downwards and backwards. The upper third of the rectum has peritoneum on its front and sides but the middle third has peritoneum only on its anterior surface. The last third of the rectum has no peritoneal covering at all since, in the male, peritoneum is reflected forwards on to the bladder, and in the female it is reflected on to the posterior wall of the vagina.

The **anal canal** starts as the bowel begins to pass through the floor of the pelvis (Fig. 8.14). The **puborectalis** muscle loops around the rectum here and pulls the proximal part of the anal canal forwards. The mucosa of the upper part of the anal canal is thrown into longitudinal **anal columns** by the terminal branches of the superior rectal artery which lie within the columns. There is also an **internal rectal venous plexus** beneath the mucosa here. The base of the columns are joined together by crescentic folds called the **anal valves** and together they form a circular **pectinate line** within the anal canal. Above the level of this pectinate line, the blood supply to the rectum comes from the **superior rectal artery**, a branch of the inferior mesenteric artery (Fig. 8.15). Below the level of the pectinate line the blood supply to the rectum comes from the **inferior rectal arteries** which are branches of the **internal pudendal arteries.** (The **middle rectal arteries** from the internal iliac arteries are said to supply only the muscle layers of the rectum.)

Likewise, the venous drainage of the rectum above the pectinate line is to the internal venous plexus and

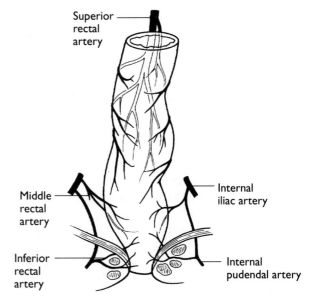

**Figure 8.15** Blood supply to the rectum comes from three sources: the superior rectal artery, the middle rectal arteries and the inferior rectal arteries.

then onwards to the superior rectal veins. These then drain into the inferior mesenteric vein and finally into the splenic vein and portal vein. Below the level of the pectinate line the venous blood collects in an **external venous plexus** which then drains to the **inferior rectal**

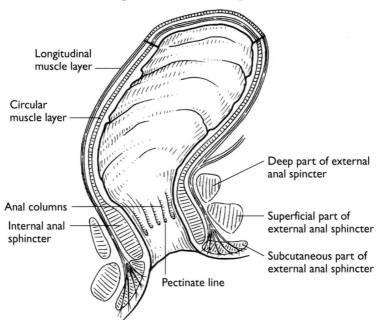

**Figure 8.14** Crescentic anal valves at the base of the anal columns form the pectinate line within the anal canal. The internal anal sphincter is the lowermost portion of the involuntary circular smooth muscle within the wall of the anal canal. The external anal sphincter consists of three rings of voluntary striated muscle fibres that run around the lower part of the anal canal.

veins and onwards to the **internal pudendal veins** and internal iliac veins. (The middle rectal veins drain only the muscle layers of the rectum.) Prolonged distension of the rectal veins (perhaps because we are upright most of the time and there are no valves in the inferior mesenteric vein) may give rise to **haemorrhoids** (sometimes called piles).

The thickened distal circular smooth muscle fibres of the bowel form the **internal anal sphincter**. These muscle fibres are supplied by involuntary autonomic nerve fibres. There is also a strong voluntary striated muscular sphincter called the **external anal sphincter**. There are three circular groups of muscle which together make up the external anal sphincter. One group, or circular band of muscle fibres, runs from the coccyx posteriorly around the anal canal and on to the perineal body anteriorly. This is called the **superficial part** of the external anal sphincter and it is important to remember it has a bony attachment. Rather confusingly, a **subcutaneous part** of the external anal sphincter lies beneath this level but has no bony attachment. A **deep part** of the external anal sphincter, which has no bony attachment either, also encircles the upper end of the anal canal. Just within the anal canal there is an **intersphincteric line** which can be palpated. This marks the junction under the mucosa of the superficial part of the external anal sphincter with the subcutaneous part of the external anal sphincter.

The external anal sphincter muscle and the mucosa below the level of the pectinate line are supplied by the inferior rectal branch of the pudendal nerve. The lower part of the anal canal is extremely sensitive. The nerve supply to the mucosa of the upper part of the anal canal, above the pectinate line, and to the internal anal sphincter is from autonomic plexuses. The internal anal sphincter contracts with sympathetic stimulation and relaxes with parasympathetic stimulation. The mucosa of the anal canal is sensitive only to stretch and it is distension above the pectinate line that initiates a defaecation reflex.

## The portal vein

The portal vein drains the whole gastrointestinal tract (Fig. 8.16). The portal vein is formed by the splenic vein, the superior mesenteric vein and the inferior mesenteric vein. In addition there are the smaller tributaries, the left and right gastric veins (which you

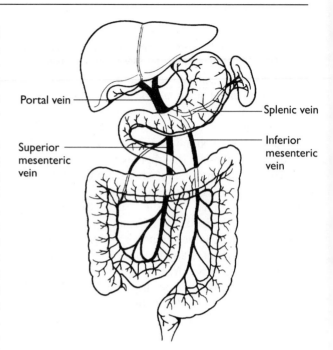

**Figure 8.16**  The portal vein forms behind the pancreas through fusion of the splenic, superior and inferior mesenteric arteries.

will remember collect some venous blood from the oesophagus) and the cystic vein from the gall bladder. The portal vein forms behind the pancreas by fusion of the splenic vein and superior mesenteric vein. It then passes up behind the pancreas within the free margin of the lesser omentum to the porta hepatis of the liver.

## Applied anatomy of the abdomen

Look first at the radiograph of a 'barium meal' (Fig. 8.17). Identify part of the fundus of the stomach which contains air, the incisura angularis, the antrum of the stomach, pyloric sphincter and the so-called **duodenal cap**. The pyloric sphincter is muscular and surrounds a thin pyloric canal that runs into the duodenum. The duodenal cap is conical with a well defined base. It shows in this way on a radiograph because the first part of the duodenum runs backwards and because the lumen of the first part is smooth inside and held open by the pylorus of the stomach. The rest of the duodenum is thrown into folds called plicae circulares. Rugae in the pyloric region usually follow the direction of food flow through the stomach. Where there is disease, such as

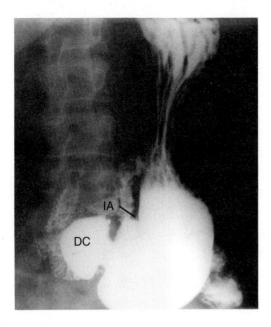

**Figure 8.17**   Radiograph of a barium meal showing the fundus, rugae, incisura angularis (IA), the narrow pyloric sphincter and the duodenal cap (DC).

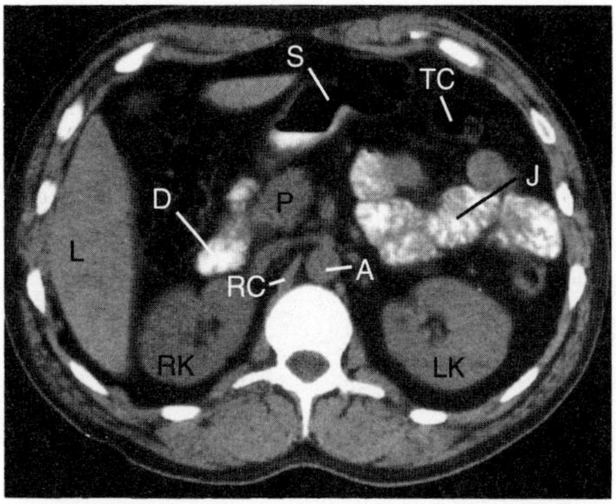

**Figure 8.18**   Computed tomogram of the abdominal cavity at the level of L1. The right lobe of the liver (L), the right kidney (RK) and the left kidney (LK) are clearly visible at this level. The abdominal aorta (A) lies in front of the vertebral body of L1. RC is the right crus of the diaphragm. There is contrast medium in the duodenum (D) in front of the right kidney and in the jejunum (J). The head of the pancreas (P) lies against the duodenum. Air is visible in the fundus of the stomach (S) and the transverse colon (TC).

a gastric ulcer, their pattern often becomes altered and they may then, for example, radiate out from the centre of the ulcer.

Duodenal ulcers result when the mucosa is eroded to form a crater-like depression in the wall of the duodenum. Recall that the gastroduodenal artery is a posterior relation of the duodenum. Erosion through the posterior wall of the duodenum can therefore easily result in severe haemorrhage into the peritoneal cavity.

Figure 8.18 is a computed tomogram (CT) of the abdominal cavity at the level of L1. The posterior abdominal wall is to the bottom of the image and the right side of the patient is to the left of the image since convention holds that CT scans are viewed as if from below. Use the key in the legend for Figure 8.18 to identify the structures and their relationships at this level in the abdominal cavity.

At the end of this volume in Chapter 10 there is a summary overview of autonomic nerves in the abdomen and pelvis. Before that we will describe how and where pain from the stomach, appendix and ureters commonly presents. In general, as we saw in the thorax, pain from visceral structures radiates out from the same segment of the spinal cord into which the visceral afferent fibres run. Afferent fibres from the stomach run back to segments T7 and T8 along the

same path of the sympathetic motor nerves in the greater splanchnic nerve. Pain is therefore interpreted by the brain to be over the dermatome of T6 and T7, that is, just below the sternum.

Early pain from the appendix passes along afferent nerves to the dorsal roots of spinal cord segment T10. It is therefore felt over the umbilical region. As inflammation of the appendix worsens, the peritoneum overlying the appendix in the lower right quadrant of the abdomen is increasingly irritated. Parietal peritoneum is supplied by somatic segmental nerves and is extremely sensitive, especially to stretch, and so the pain becomes localized to the region overlying the appendix. The pain of gently pressing in this region with the fingers is bad but that of **rebound tenderness** on suddenly removing them, as the peritoneum recoils, is considerably worse.

The ureters are supplied with afferent pain fibres that return to segments T12 and L1 with the lowermost splanchnic nerves. Stones in the ureters give rise to extreme sharp stabbing colic as they pass down from the kidney to the bladder. Pain from the ureters is referred to the areas of skin supplied by T12 and L1. This begins in the back between the ribs and iliac crest and then radiates down to the inguinal region and even the scrotum.

# THE PELVIS AND PERINEUM

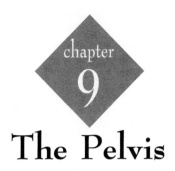

chapter
9

# The Pelvis

The bony pelvis provides support for the abdominal and pelvic viscera and is modified in the female to meet the requirements of childbirth. It also transmits the weight of the trunk from the vertebral column to the femoral heads. It follows that some details of the bony pelvis are best studied with the vertebral column and some with the lower limb. These are described in Volume 1 of *Core Anatomy For Students*. In this chapter, however, we will concentrate on the contents of the pelvis. In the next chapter we will describe the structures that lie in the perineum below the pelvic diaphragm. To do this we need to look at the bony framework of the pelvic cavity in more detail and then consider some of the differences between the male and female pelves.

## The bones of the pelvis

Look first at Figure 9.1. The two hip bones, or **innominate bones**, articulate with the **sacrum** at the sacro-

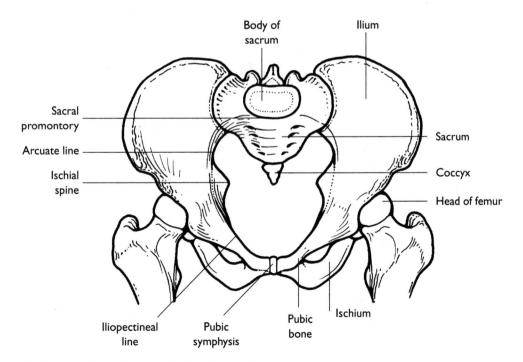

*Figure 9.1* The bones and major bony landmarks of the pelvis.

iliac joints. These are synovial joints. At the tip of the sacrum a few fused vertebral segments form a separate bone called the **coccyx**. In the midline anteriorly, the innominate bones articulate with each other at the pubic symphysis. You will remember that the pubic symphysis is one of the midline secondary cartilaginous joints. The articulated pelvis as a whole resembles a basin with a wide **superior aperture** at the top and a narrower **outlet** below.

The sacrum is formed from five fused segments of the vertebral column. The body of the first sacral segment bears the weight of the trunk which is then passed bilaterally through each **ala** or **lateral mass** of the sacrum to the sacroiliac joints. The **sacral promontory** is the most forward point on the superior margin of the body of the first sacral segment. The **arcuate lines** pass anteriorly from the sacral promontory and run into the **iliopectineal lines**, which lead to the pubic crest and then to the midline anteriorly. The continuous bony **pelvic brim** defined by these lines is known as the **pelvic inlet**. The **greater pelvis** lies above the plane of the pelvic inlet and the **lesser** or **true** pelvis lies below.

Each innominate bone is made up of three separate bones which fuse together. In children these bones are separated by a 'Y-shaped' or **triradiate** growth plate that runs through the acetabulum of the hip joint (Fig. 9.2). The triradiate cartilage fuses during puberty. The most superior of the three bones is called the **ilium**. The ilium is surmounted by the **iliac crest**

which runs from the **posterior superior iliac spine** to the **anterior superior iliac spine** over the **ala** of the iliac bone. The **iliac fossa** forms the inner aspect of the iliac bone and is completely covered by the iliacus muscle in life. The superior part of the acetabulum is part of the ilium and this is the weight-bearing portion of the joint socket.

The **pubic bone** has superior and inferior **pubic rami** which meet each other at the **body** of the pubic bone anteriorly. The bodies of the right and left pubic bones join together at the **pubic symphysis**. We have studied the pubic crest, pubic tubercle and pectineal line of the pubic bone in Chapter 6 along with the details of the inguinal region, but identify these features again in the diagrams of the bony pelvis. They form part of the superior aperture of the pelvis (Figs 9.1 and 9.3).

The third bone of the innominate is called the **ischium**. The ischium forms the posterior third of the acetabulum as defined by the triradiate cartilage. The part adjoining the ilium is the body of the ischium. Its posterior border is divided into a **lesser sciatic notch** and a **greater sciatic notch** by the **ischial spine** which projects between the two notches (Fig. 9.3). Strong ligaments pass between the spine of the ischium and the sacrum, and between the ischial tuberosity and the sacrum (Fig. 9.4). These are called the **sacrospinous** and **sacrotuberous** ligaments respectively. They transform the lesser and greater sciatic notches into the **lesser** and **greater sciatic foramina**. The **ischial tuberosity** is the part of the pelvis

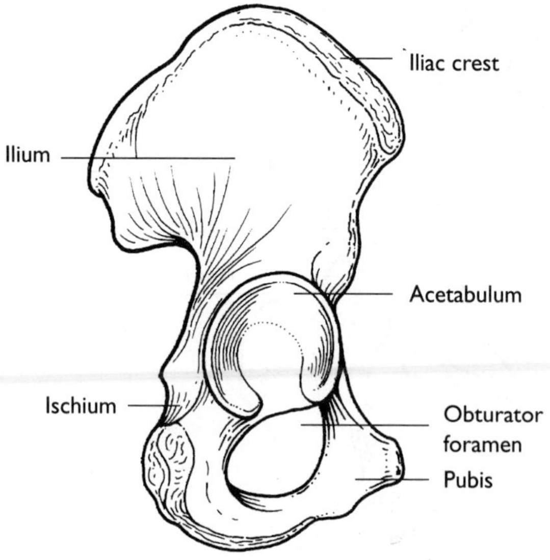

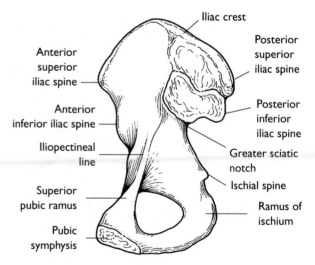

**Figure 9.2**  The innominate bone is made up of the ilium, pubis and ischium. These fuse at the triradiate cartilage during puberty.

**Figure 9.3**  Important bony landmarks of the innominate seen from the medial aspect of the bone when disarticulated.

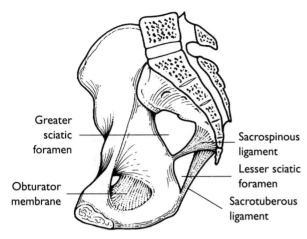

**Figure 9.4** The sacrospinous and sacrotuberous ligaments form the greater and lesser sciatic foramina together with the sacrum, ilium and ischium. The obturator membrane closes over the obturator foramen in life.

we sit on. The tuberosity is roughened and curves round to become the posterior border of the ischium. The **ramus of the ischium** sweeps round from the acetabulum to join up with the inferior pubic ramus. The **obturator foramen** of the pelvis is bounded by the superior and inferior pubic rami and by the ischium posteriorly and by the ramus of the ischium, which fuses with the inferior pubic ramus. In life the obturator foramen is covered over by the **obturator membrane** (Fig. 9.4) and the **internal** and **external obturator muscles** on either side of it. The **ischiopubic ramus** runs from the pubic symphysis in the midline anteriorly to the ischial tuberosity behind. The angle formed by the right and left ischiopubic rami is called the **subpubic angle**.

## The male and female bony pelvis

The inlet of the male pelvis is narrowed by the large acetabulae of the hip joints at the sides and by the sacral promontory protruding in from behind. This gives the pelvic inlet of the male a rather 'heart-shaped' appearance when viewed from above.

In the female pelvis the hip joints are smaller and the acetabulae are relatively further apart so that the inlet is wider. You can judge this by noting that the diameter of the male acetabulum is approximately equal to the length of the superior pubic ramus when measured from the anterior margin of the acetabulum to the pubic symphysis (Fig. 9.5). This distance is greater than the diameter of the acetabulum in the female pelvis. The sacrum as a whole is positioned further back in the female pelvis (Fig. 9.6). The alae of the sacrum are also wider in the female so that the body of the first sacral segment makes up less of the total width of the sacrum in the female than in the male.

The pelvic outlet in the male pelvis is smaller than that of the female pelvis because of the forward position of the sacrum and because the ischial spines protrude more medially into the outlet. In the female pelvis the sacrum is 'displaced' backwards so that, when you compare a male and female pelvis from the side, the greater sciatic notch looks opened out (Fig. 9.6). The male notch looks like an inverted letter 'J' superiorly; the female notch looks more like an inverted letter 'L'. The subpubic angle in the male pelvis is about 60–70°. In a female pelvis it is 90° or greater. Thus the ischial tuberosities and the ischial spines are positioned more laterally in the female pelvis. One final point is that the true pelvis of the female (below the pelvic brim) is not as tall as the male pelvis.

In summary, the female pelvic inlet is wider than it is long. In contrast to the inlet, however, the female pelvic outlet is longer anteroposteriorly than it is wide (Fig. 9.7). The increased transverse diameter of the female pelvic inlet and increased anteroposterior diameter of the female pelvic outlet are essential to allow normal childbirth. During childbirth the infant's head first engages with its maximum diameter lying across the pelvic inlet. As it passes towards the pelvic outlet the head changes its orientation and rotates through 90° so that the face is turned pos-

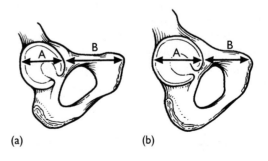

(a)                    (b)

**Figure 9.5** The diameter of the acetabulum (A) is approximately equal to the length of the superior pubic ramus (B) in the male pelvis (b). In the female pelvis (a) the acetabulum is smaller and the superior pubic ramus longer, such that A and B are no longer equal in length.

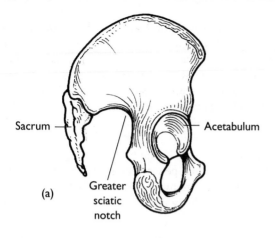

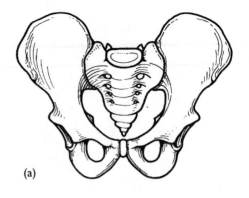

Labels for figure 9.6: Sacrum, Acetabulum, Greater sciatic notch, (a), (b)

Labels for figure 9.7: (a), (b)

**Figure 9.6**  The greater sciatic notch in the female pelvis (a) is drawn out posteriorly such that it resembles an inverted letter 'L'. The male notch (b) is smaller and resembles an inverted letter 'J' more.

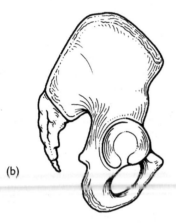

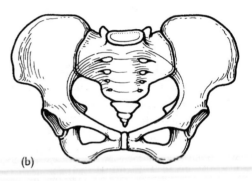

**Figure 9.7**  The female pelvis (b) is less tall than the male pelvis (a). The female pelvis has a wider sacrum but one with a smaller sacral body. The female pelvis has a wider pelvic inlet. The sub-pubic angle is also greater in the female than in the male.

teriorly. In this position the maximum diameter of the head now lies in an anteroposterior orientation as it passes through the pelvic outlet at birth. The birth canal of the female pelvis is therefore adapted to allow these events to occur as easily as possible during childbirth.

## The muscular walls and floor of the pelvis

When we stand upright, the anterior superior iliac spines of the pelvis lie in the same vertical plane as the body of the pubic bone below. In this position the posterior aspect of the body of the pubic bone, the pubic rami and the blocked off obturator foramen provide some support for the pelvic and abdominal viscera above. However, if it were not for the presence of a muscular diaphragm that fills in the gap between the pubic bone anteriorly and the coccyx and sacrum behind, the contents of the pelvis would tend simply to fall to the floor.

Before we describe the muscles of the pelvic diaphragm we need to mention two muscles that form the walls of the pelvis and are really best studied in detail with the lower limb. These are the **obturator internus** and the **piriformis** (Fig. 9.8). Obturator internus arises from the bone surrounding the obturator foramen inside the pelvis and from the obturator membrane which covers over the obturator foramen itself. Obturator internus then passes posteriorly, converging to a tendon at the margin of the *lesser* sciatic

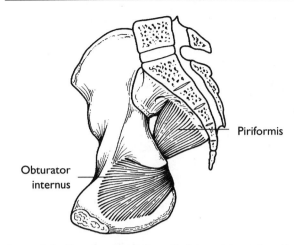

**Figure 9.8** Two muscles of the wall of the pelvis are the obturator internus, which passes out of the lesser sciatic foramen, and the piriformis, which passes out of the greater sciatic foramen. Both muscles are lateral rotators of the hip.

foramen. Here the muscle turns through 90° and runs laterally out of the pelvis and into the greater trochanter of the femur. When it contracts it rotates the thigh laterally. Within the pelvis the muscle is covered with a thickened fascia.

The piriformis arises from the middle three segments of the sacrum. It is arranged in such a way to allow nerves free passage out of the anterior sacral foramina and into the pelvic cavity. Later, we will see that the sacral plexus of nerves forms on its anterior surface. Piriformis passes laterally out of the pelvis through the *greater* sciatic foramen. It converges to a short tendon which inserts on to the posterior part of the greater trochanter of the femur. Like obturator internus it is a lateral rotator of the thigh. (It contains many muscle spindles and is probably very important

in relaying proprioceptive information back to the spinal cord.) The obturator internus and the piriformis muscles wall off the sides and back of the true bony pelvis respectively.

Now we are in a position to concentrate further on two important muscles of the pelvic floor called **levator ani** and **coccygeus**. It is these muscles that form the floor of the pelvis or, as it is sometimes known, the **pelvic diaphragm** (Fig. 9.9). The pelvic diaphragm is convex *downwards*, unlike the diaphragm between the thorax and abdomen. It curves markedly from side to side. It is covered on its superior surface by an extension of the transversalis fascia which within the pelvis is simply known as the **superior fascia of the pelvic diaphragm**. The rectum, the vagina in the female, and the urethra all have to pass through the pelvic diaphragm.

The levator ani takes origin from the inner surface of the pubic bone near the lower margin of the pubic symphysis (Fig. 9.9). It then attaches to a line of thickened fascia that runs from front to back over the obturator internus muscle called the **arcus tendineus**. Finally, the line of origin of levator ani runs on to the ischial spine. All of the muscle fibres of levator ani are destined to swing in towards the midline and fuse with muscle fibres from the other side to form a midline raphé or tendinous seam (Fig. 9.10). However, there is a gap in the midline immediately behind the body of the pubic bone called the **urogenital hiatus**. The urogenital hiatus gives passage to the urethra and, in the female pelvis, to the vagina. There is also another gap in the midline further back which gives passage to the anal canal. As a result, the raphé is divided into a small section between the rectum and the urogenital hiatus which is tough and fibrous in

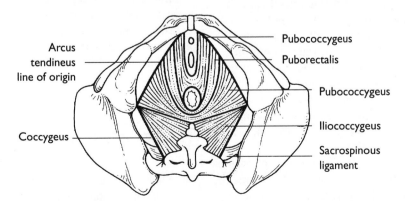

Arcus tendineus line of origin

Coccygeus

Pubococcygeus

Puborectalis

Pubococcygeus

Iliococcygeus

Sacrospinous ligament

**Figure 9.9** The levator ani, or pelvic diaphragm, is made up of several groups of muscle fibres which are seen from below here. The anal canal passes through the pelvic diaphragm, and the urogenital hiatus transmits the urethra and the vagina in the female pelvis.

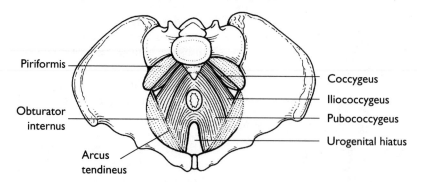

**Figure 9.10** Seen from above, the muscles that make up the walls and floor of the pelvis are now all visible. The arcus tendineus overlying obturator internus is the line of origin of much of levator ani.

nature and is called the **perineal body** and a longer section behind the rectum called the **anococcygeal raphé**.

Levator ani can be divided up into smaller portions from front to back and this may make its functions easier to understand. It does not, however, mean there is great merit in learning all the separate parts. The fibres that arise from the body of the pubic bone pass around the margins of the urogenital hiatus and insert into the perineal body. This part of the muscle is known as the **pubococcygeus muscle**. Some of its fibres may run into the fascia covering the vagina or prostate gland. Some muscle fibres of pubococcygeus pass behind, or form a sling around, the vagina and form a potential sphincter about one-third of the way up the length of the vagina. The fibres of levator ani that arise more laterally from the pubic bone and from the first part of the arcus tendineus constitute the **puborectalis muscle**. These fibres insert just behind the rectum into the anococcygeal raphé (Fig. 9.11). The most anterior fibres of puborectalis from the left and right form a **puborectal sling** that pulls the rectum forwards towards the pubis. This is an important mechanism for maintaining rectal continence. The fibres of levator ani that arise from the most posterior part of the arcus tendineus and from the ischial spine and then insert into the most posterior part of the anococcygeal raphé constitute the **iliococcygeus muscle**. In truth, these are feeble muscle fibres, especially in the elderly, and are often little more than fibrous in nature.

**Coccygeus** (or more correctly ischiococcygeus since it runs from the ischial spine to the coccyx) forms the most posterior part of the pelvic diaphragm. Coccygeus is one of the muscles that wags the tail in many animals. In humans, however, it runs between two

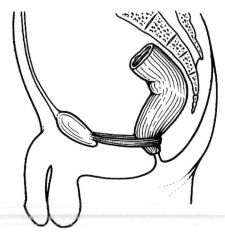

**Figure 9.11** Puborectalis is a part of levator ani. Its fibres sweep around the lower part of the rectum and play an important part in the mechanism of anal continence.

bony points which are now immobile and because of this much of it has regressed to a ligament called the sacrospinous ligament which we have already described. The coccygeus muscle lies in the coronal plane and together with the piriformis forms the muscular posterior wall of the pelvic cavity.

# The urinary bladder and the ureters

The bladder lies in the anterior part of the true pelvis and rests on the posterior surface of the bodies of the pubic bones, above the urogenital hiatus. The sides of the two pubic bones mould the front of the bladder so that it resembles the front of a boat (Fig. 9.12). The bladder also has a domed triangular-shaped **superior surface**. The flat **posterior surface** is called the **base** of the bladder. The superior surface is in contact with the mobile abdominal viscera through parietal peritoneum, and the posterior surface, or base, relates either to the vagina and uterus in the female or to the rectum in the male (see below).

The bladder is an extraperitoneal structure like the ureters and kidneys. During embryonic life the anterosuperior margin of the bladder is joined to a tubular duct which runs upwards in the anterior extraperitoneal space to reach the umbilical cord. This duct is the **urachus**. After birth the urachus degenerates into the **median umbilical ligament** which can still be identified in a peritoneal fold running from the bladder to the umbilicus called the **median umbilical fold**. The bladder wall contains smooth muscle fibres called the **detrusor muscle**. When the bladder is empty the interior is thrown into folds but when it is distended it appears quite smooth. When the bladder distends it expands up, in front of the peritoneal cavity (Fig. 9.12). Occasionally, when severely distended, it may reach as far up as the umbilicus. Internally, the opening of the ureters into the base of the bladder and

the entrance into the urethra define a triangular area known as the **trigone**. The wall of the trigone does not expand and contract like the wall of the rest of the bladder and always appears smooth in life rather than folded like the rest of the bladder (Fig. 9.13). The trigone is highly sensitive to pain. Around the internal urethral orifice there is a ring of smooth muscle known as the **internal sphincter urethrae**. This sphincter is relaxed by parasympathetic nerve activity and closed by sympathetic nerve activity.

## The ureters in the pelvis

We last traced the ureters over the posterior abdominal wall to the pelvic brim where they crossed the bifurcation of the common iliac arteries (Fig. 9.14). From here the ureters run down into the true pelvis and cross over the internal iliac vessels and the obturator vessels and nerve. At the pelvic floor the ureters run forward to enter the bladder at its base or posterior surface. At this point the ductus deferens in the male hooks over the top of the ureters.

In the female the uterine artery runs over the top of the ureter as it runs beneath the broad ligament. This relationship is a very important relationship to visualize and understand because the uterine artery must be distinguished from the ureter at hysterectomy and the artery, not the ureter, clamped off.

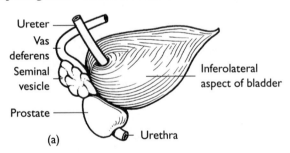

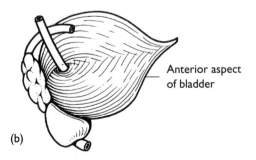

**Figure 9.12** The bladder lies between the two pubic bones anteriorly and is shaped accordingly here. The superior surface becomes increasingly rounded as the bladder fills (b) and expands upwards.

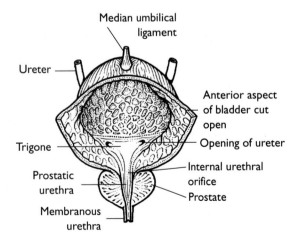

**Figure 9.13** The ureters pass obliquely through the bladder wall to open at the trigone. The urethra leaves the bladder at the base of the trigone. (After Sabotta J and McMurrich JP. *Atlas of Human Anatomy.* New York: GE Stechert, 1930.)

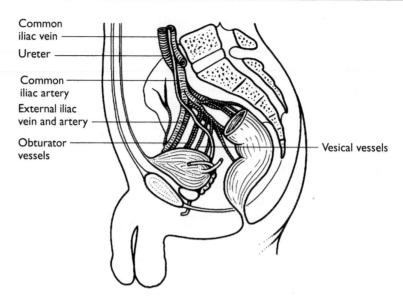

**Figure 9.14** Branches of the internal iliac artery and internal iliac vein run across the lateral walls of the true pelvis to the pelvic viscera. The ureters cross these vessels and run downwards towards the bladder.

The ureter lies close to the cervix here as the uterine artery crosses it.

At the base of the bladder the ureters then run obliquely downwards through the muscle of the bladder wall to open lower on the posterior wall at the top of the trigone. The ureters have both a sympathetic and parasympathetic nerve supply. Urine is moved to the bladder by peristaltic action of the smooth muscle in the wall of the ureters.

# The relations of the male bladder

In the male, the neck of the bladder lies on, and is intimately associated with, the superior surface of the **prostate gland** (Figs 9.12 and 9.15). The bladder and prostate gland lie anterior to the rectum, and the prostate and posterior wall of the bladder can be palpated *per rectum* through the anterior wall of the rectum as far as the peritoneum of the **rectovesical pouch** extends downwards. The urethra passes out of the bladder and through the substance of the prostate gland. The neck of the bladder and the prostate are fixed to the body of the pubic bone by **puboprostatic ligaments.**

The vas deferens leaves the deep inguinal ring and runs downwards into the pelvis towards the superior border of the posterior wall of the bladder. Each vas

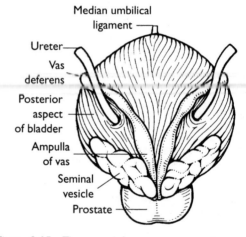

**Figure 9.15** The vas deferens runs over the ureter as it approaches the midline of the bladder posteriorly. The seminal vesicles lie in the groove between the bladder and the prostate gland posteriorly.

deferens runs over the top of the ureter near the superior border of the bladder and approaches the midline. Here, it turns down the posterior wall of the bladder. Either side of the midline on the posterior surface of the bladder and above the base of the prostate gland each vas deferens widens into an **ampulla**. Lateral to the ampullae in the groove between the bladder and the prostate gland are the **seminal vesicles.** Ducts from the seminal vesicles join the ampullae of each vasa deferentia to form the right and left **ejaculatory ducts**. The two ejaculatory ducts then run through the substance of the prostate gland and open obliquely on to the **seminal colliculus** of the prostatic urethra, lateral to the **prostatic utricle** (see below).

# The prostate gland

The prostate lies below the bladder and surrounds the proximal part of the urethra. It is conical in shape with its base adjacent to the bladder. The prostate is normally palpable *per rectum* and feels hard but 'elastic'. A full bladder displaces the prostate downwards and makes it easier to feel. The whole gland is encapsulated by a fibrous sheath and there is a **prostatic venous plexus** in this sheath anteriorly and at the sides of the prostate. The ejaculatory ducts divide the prostate into an upper and lower part. The upper part is sometimes called the **median lobe**. The lower part that lies behind the prostatic urethra can be divided into **right** and **left lateral lobes**. The part of the gland in front of the prostatic urethra is called the **anterior lobe**. The lobes are really useful only for descriptive purposes and they do not demarcate regions of the gland histologically. The ducts from the prostate gland are 20 or so in number and open into the prostatic urethra. We will describe these in further detail below when we consider the male urethra. The prostate enlarges both at puberty and then again in adulthood. If it enlarges a lot it can obstruct the urethra at the neck of the bladder. Sixty per cent of men have cancer cells in their prostate at post mortem and prostatic cancer is common. Spread of prostatic cancer is unfortunately easy since the prostatic venous plexus communicates freely with the vertebral venous plexus. Straining to urinate when the prostate is enlarged may actually force venous blood from the prostatic venous plexus into the vertebral venous plexus.

# The relations of the female bladder

In the female pelvis, the bladder lies directly over the urogenital hiatus since there is no prostate gland beneath it. The bladder is fixed to the body of the

pubic bone by the **pubovesical ligaments**. The posterior surface of the bladder in the female is related to the upper part of the anterior wall of the vagina and also to the cervix and uterus. Peritoneum runs off the front of the uterus where the cervix joins the body of the uterus and on to the surface of the bladder (Fig. 9.16). There is here a **uterovesical pouch** of peritoneum between the bladder and uterus.

# The uterus, vagina and ovaries

It is essential, eventually, to describe the peritoneal coverings of the female pelvic organs in some detail after we have described the basic anatomy of the female reproductive organs. Before doing this, however, it is a good idea to describe the general arrangement of the pelvic viscera and peritoneum.

Understand that the uterus projects upwards and forwards into the pelvic cavity in the midline, behind the bladder (Fig. 9.17). The uterine tubes extend

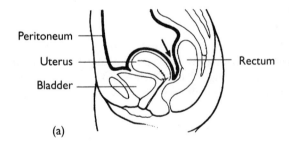

(a)

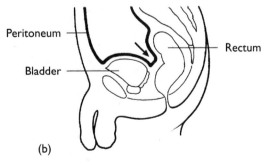

(b)

**Figure 9.16**   In the female pelvis (a), peritoneum forms a rectouterine pouch as it runs down over the rectum and then up over the uterus in the midline. In the male pelvis (b) peritoneum forms a rectovesical pouch between the rectum and bladder.

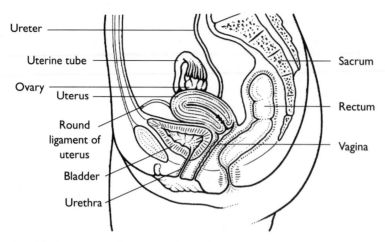

**Figure 9.17** In the female pelvis the uterus projects upwards and forwards in the midline. The uterine tubes pass laterally from the uterus towards the pelvic wall.

laterally from the fundus of the uterus like outstretched arms reaching towards the pelvic brim. The peritoneum lining the female pelvic cavity is draped over the bladder anteriorly and the uterus and uterine tubes immediately behind and over the uppermost part of the rectum posteriorly. We have already described the uterovesical pouch of peritoneum between the bladder and uterus where peritoneum extends a little way down between these organs. There is also a **rectouterine pouch** of peritoneum (or **pouch of Douglas**) between the uterus and rectum posteriorly (Fig. 9.16). Obviously this does not exist in the male and there is simply a **rectovesical pouch** of peritoneum between the bladder and rectum posteriorly. The rectouterine pouch is the most inferior

point of the peritoneal reflection in the female pelvis and it is important to realize that blood, pus or any free-floating material within the peritoneal cavity will easily collect in this pouch in patients lying in the supine position. Study Figure 9.16 and understand how samples of its contents can be drawn through a needle via the posterior fornix of the vagina.

Lateral to the uterus in the female pelvic cavity, peritoneum drapes over the uterine tubes exactly like a sheet hanging over a washing line (Fig. 9.18). It falls into the floor of the pelvis as a double-layered fold of peritoneum and then splits into two again as it spreads round, up and out of the pelvis anteriorly and posteriorly. The double fold of peritoneum that hangs over and beneath the uterine tubes is known as

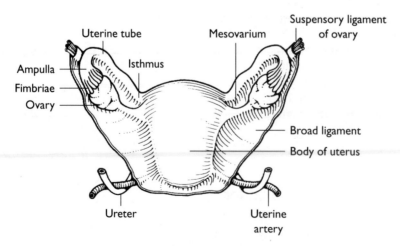

**Figure 9.18** The broad ligament hangs over the uterine tubes and attaches to the walls of the pelvis. Ovarian vessels run through the suspensory ligament of the ovary. Notice, in this view from behind, the important relationship of the uterine artery to the ureters.

the **broad ligament of the uterus**. The broad ligaments extend laterally to the pelvic brim and attach a little way down the lateral wall of the true pelvis and on to the pelvic diaphragm. This attachment of the broad ligament is called the **root** of the broad ligament. The broad ligament helps to suspend the uterus and uterine tubes across the true pelvic cavity. The broad ligaments divide the female pelvic cavity into an **anteroinferior compartment** and a **posterosuperior compartment**. The bladder lies in the anteroinferior compartment and the rectum lies in the posterosuperior compartment. With this general outline clear we are now in a position to study the anatomy of the uterus, vagina, uterine tubes and ovaries.

## The uterus

The uterus has a **fundus**, **body** and a **cervix**, or neck (Fig. 9.19). The body of the uterus extends up to the level at which the uterine tubes enter it and above this level the body becomes the fundus of the uterus. (A fundus of any cavity is that part farthest from the entrance into the cavity. Fundus: L = the bottom of a bag.) There is a bend between the cervix and the body of the uterus so that usually, the body of the

uterus lies more anterior than the cervix (Fig. 9.20). This bend is called **uterine anteflexion**. Furthermore, the axis of the cervix and the axis of the vagina form an angle of about 90°. This is called **anteversion of the uterus**. A distended bladder or rectum will alter the degree to which the uterus falls forwards.

The uterus has very thick fibromuscular walls. The lumen of the uterus opens inferiorly first through an **isthmus**, then through the **cervical canal** and finally into the vagina at the **external uterine os**. The lower end of the uterus protrudes slightly into the upper part of the vagina so that a small rim of the vaginal lumen encircles the cervix. The circular recess that results surrounds the cervix and is known as the **fornix** of the vagina. This can be divided for descriptive purposes into the anterior, posterior or lateral parts of the fornix (Figs 9.17 and 9.19).

The cervix of the uterus and the fornix of the vagina are attached to the floor and walls of the pelvis by fibromuscular thickenings. These may help to stabilize the uterus and vagina. The **transverse cervical**

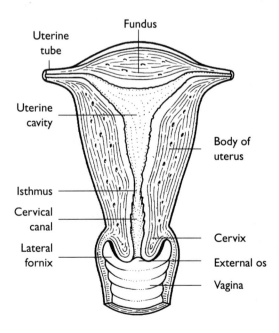

**Figure 9.19** The uterus has a fundus, a body and a cervix. The cervical canal leads via the external os into the vagina.

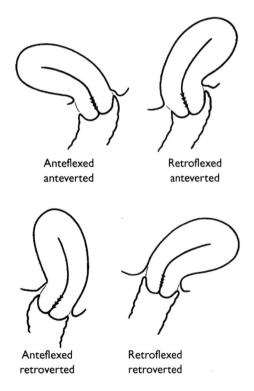

**Figure 9.20** The uterus is usually anteverted with respect to the vagina. It is usually also anteflexed at the junction of the body and cervix. All combinations of anteflexion and anteversion present clinically though. (After Ellis H. *Clinical Anatomy*. Oxford: Blackwell Scientific Publications, 1977.)

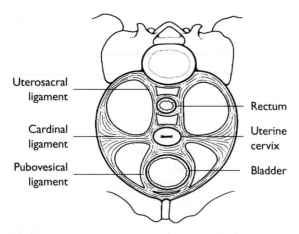

**Figure 9.21**    Fibrous ligaments tie the cervix of the uterus to the lateral and posterior walls of the pelvis. The cardinal ligaments run in the base of the broad ligament. Uterosacral ligaments run to the sacrum on either side of the rectum.

or **cardinal ligaments** run in the base of the broad ligaments towards the lateral walls of the pelvis (Fig. 9.21). Other ligaments run posteriorly from the cervix to the sacrum, either side of the rectum, in the lateral folds of the rectouterine pouch. These are called the **uterosacral ligaments**. Yet others run forwards from the uterus around the bladder to tie in with the pubovesical ligaments behind the pubic symphysis.

# The vagina

The long axis of the vagina lies at 90° to that of the uterus. The cervix, therefore, really projects into the anterior wall of the vagina. The anterior wall of the vagina is related to the bladder through loose connective tissue and lower down to the urethra, to which it is, however, firmly adherent. The posterior wall of the vagina is longer than the anterior wall. Superiorly, it is related to the rectouterine pouch which lies behind to the posterior fornix. Below the level of the rectouterine pouch the vagina is related to the rectum. Recall that about one-third of the way up the vagina some of the fibres of levator ani loop behind the vagina and create a partial sphincter around it.

# The uterine tubes

The uterine tubes pass out of the uterus at its superolateral corners (Fig. 9.22). Initially, the tube is narrow and is referred to as the **isthmus**. More laterally the tube widens into an **ampulla** which expands even more into an **infundibulum** just before it opens into the peritoneal cavity. The opening of the infundibulum is called the **abdominal ostium**. The wall of the

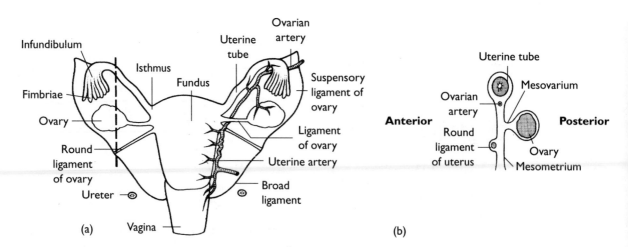

**Figure 9.22**    The ovarian artery approaches the ovary within the suspensory ligament of the ovary. The uterine artery approaches the cervix in the base of the broad ligament. The uterine tube, artery, ovary and round ligament of the ovary are all enclosed by the broad ligament.

infundibulum has a fringe of **fimbriae** which are able to 'claw' at the ovary during ovulation. These fimbriae 'wave backwards' from the lateral extremities of the uterine tubes into the posterosuperior part of the pelvic compartment.

## The ovaries

Each **ovary** lies against the wall of the pelvis at a point that roughly corresponds to the position of the acetabulum on the outside. The ovary is shaped, for want of a better description, like a prune. The upper pole is closely associated with the fimbriae of the uterine tube. The lower pole of the ovary is connected to the uterus via a fibromuscular cord which runs to a point on the body of the uterus just beneath the origin of the uterine tube. This cord is called the **ligament of the ovary**. A continuation of this cord travels onwards from the body of the uterus, back through the broad ligament towards the inguinal canal and then through the canal and into the labia majora. This continuation of the cord is known as the **round ligament of the uterus**. Both the ligament of the ovary and the round ligament of the uterus are remnants of the gubernaculum that drew the gonad into the pelvis during embryonic life (and in the case of the male onwards into the scrotum).

The ovary is covered by a layer of cuboidal epithelial cells which are continuous with the peritoneum. **Ova** within the ovary that mature and rupture though this layer are extruded into the peritoneal cavity and have to cross the peritoneal cavity to enter the uterine tube.

## The peritoneal coverings of the ovaries and uterine tubes

We have seen how the broad ligament hangs like a sheet over the uterine tubes (Fig. 9.18). Its two layers initially enclosed not only the uterine tubes but also the blood vessels to the uterus and ovary, the ligament of the ovary and the round ligament of the uterus. The outer quarter of the broad ligament that carries the ovarian artery and vein has rather confusingly been named the **suspensory ligament of the ovary** (quite distinct from the ligament of the ovary).

The infundibulum also has its own connection to the wall of the pelvis in the form of the **infundibulo-pelvic ligament**.

During development the ovary protrudes from between the two layers of the broad ligament posteriorly, pulling a visceral peritoneal covering with it (Fig. 9.22). The 'root' of this peritoneal extension or mesentery from the broad ligament is called the **mesovarium**. The part of the broad ligament above the root of the mesovarium is called the **mesosalpinx** (pertaining to the uterine tubes). The part of the broad ligament below the level of the root of the mesovarium is called the **mesometrium** (pertaining to the uterus). The round ligament of the uterus on its way to the inguinal canal raises a fold on the anterior surface of the broad ligament, even in the adult.

## The blood supply to the pelvis

The blood supply to the pelvis come from the internal iliac artery. The internal iliac artery runs down into the pelvis towards the greater sciatic notch. Just above the notch it is said to divide into **anterior** and **posterior divisions** but variations are frequent and this is hardly a useful description (Fig. 9.23). From the outset it helps to realize that many of the branches of the internal iliac artery are destined to leave the pelvic cavity. We have no need to study these arteries here but if we mention them first, briefly, it clears the way for describing the arteries that supply the pelvic organs. One of these is the **superior gluteal artery**. This leaves the pelvis through the top of the greater sciatic notch. There are other branches, to the muscles of the posterior wall of the abdomen and pelvis, but these will not be described.

Other branches of the internal iliac artery that leave the pelvis include the **obturator artery**. This squeezes out of the obturator foramen with the obturator nerve. A large trunk leaves the pelvis over the lower margin of the greater sciatic foramen. This trunk immediately divides into the **inferior gluteal artery** and the **internal pudendal artery**. Recall that the piriformis muscle also passes out of the pelvis through the greater sciatic notch. In fact, it passes out *between* the superior and inferior gluteal arteries and this makes each artery easier to identify and remember. The internal pudendal artery is the artery of the perineum and we will return to it later when we study this region.

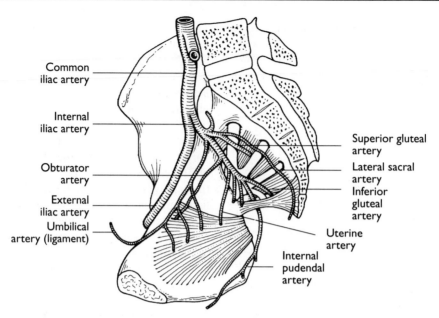

Common iliac artery

Internal iliac artery

Obturator artery

External iliac artery

Umbilical artery (ligament)

Superior gluteal artery

Lateral sacral artery

Inferior gluteal artery

Uterine artery

Internal pudendal artery

**Figure 9.23** The umbilical artery gives off the superior vesical arteries. Middle and inferior vesical arteries (or uterine and vaginal arteries in the female) arise directly from the internal iliac artery. Other branches of the internal iliac artery pass out of the pelvis.

We are now left with the branches of the internal iliac artery that supply the pelvic organs. It is far better to try to remember what they supply than to attempt to learn them off by heart. In fact, only the most important arteries will be discussed here. Three branches in the male go to supply the bladder, seminal vesicles and prostate gland. These are the **superior middle** and **inferior vesical arteries**. The **umbilical artery** gives rise to the superior vesical artery and the middle vesical artery quickly sprouts from the superior vesical branch. Their territories overlap from the superior aspect of the bladder down to the inferior part of the prostate gland. The umbilical artery needs a further mention since it continues on after giving off the superior vesical artery over the top of the bladder and then turns up to run on the inner aspect of the anterior abdominal wall. Its lumen quickly fades away but can still be recognized as the **medial umbilical ligament**.

(We have now studied three structures that raise *folds* of peritoneum on the posterior surface of the anterior abdominal wall: the median umbilical ligament, the remnant of the urachus; the medial umbilical ligament, remnant of the umbilical artery; and the inferior epigastric artery, the lateral umbilical ligament. Confusingly, the folds these structures raise are named **median, medial** and **lateral umbilical folds**.)

The **middle rectal artery** is another important

branch of the internal iliac artery and it leaves below the vesical arteries to supply muscle in the wall of the rectum. Remember we have already seen in our study of the abdomen that the *superior* rectal artery is a terminal branch of the inferior mesenteric artery and that the *inferior* rectal artery is a branch of the internal pudendal artery.

In the female, the inferior vesical artery is replaced by the **vaginal artery**. Besides supplying the vagina, the vaginal artery also supplies a part of the bladder anterior to the vagina and part of the rectum posterior to the vagina. But by far the most important artery of the pelvis to remember is the **uterine artery** of the female. The uterine artery is large. It runs down the wall of the true pelvis towards the lower border of the broad ligament. Then it turns and runs medially between the two layers of the broad ligament towards the cervix (Fig. 9.24). Within the broad ligament it runs over the top of the ureter, which is running beneath the peritoneum here towards the bladder. The uterine artery turns upwards at the cervix and runs alongside the body of the uterus. It gives branches that now travel laterally with the uterine tube, the round ligament of the uterus and the ligament of the ovary.

Now is a good time to recall that the **ovarian artery** is a branch of the abdominal aorta. It arises just below the level of the renal artery, runs down the posterior

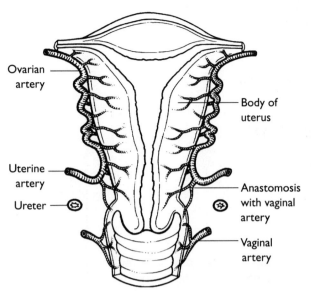

Ovarian artery

Body of uterus

Uterine artery

Ureter

Anastomosis with vaginal artery

Vaginal artery

**Figure 9.24** The ovarian, uterine and vaginal arteries anastomose with each other in the female in the same way as the vesical arteries in the male.

abdominal wall and over the brim of the pelvis to enter the suspensory ligament of the ovary. From this point it runs into the mesovarium and anastomoses with branches of the uterine artery which are passing laterally to meet it.

Each of the pelvic organs is surrounded by a plexus of veins which eventually drain into the internal iliac veins via tributaries that travel with the arteries we have described. The exceptions are the **ovarian veins** which on the left drain into the renal vein and on the right drain directly into the inferior vena cava. The pelvic venous plexuses communicate freely with the vertebral venous plexuses and also with the portal system via anastomoses between the middle and superior rectal veins. This has important implications for the spread of tumours from the pelvis.

chapter

**10**

# The Perineum

The **perineum** is the region beneath the pelvic diaphragm. It can be divided into two parts, for the purpose of description, by an imaginary line drawn across the pelvis between the left and right ischial tuberosities. The region in front of the line is known as the **urogenital triangle** (Fig. 10.1). The apex of the urogenital triangle is just behind the pubic symphysis. The region behind the line is called the **anal triangle** and it has its apex at the coccyx. For our purposes it makes sense to study the structures associated with the male and female urogenital triangle first and then return to the anatomy of the anal triangle later.

## The urogenital triangle

The lateral walls of the urogenital triangle are formed by the ischiopubic rami and the inner surface of the obturator internus muscle below the level of the arcus tendineus. Since the pelvic diaphragm is curved downwards you will realize that it drops to the same level as the ischiopubic rami in the midline, but attaches much higher than this at the arcus tendineus (Fig. 10.2). The **perineal membrane** is a flat fibrous sheet that runs between the lower borders of the left and right ischiopubic rami. To be truthful, it is a far from substantial structure and even its existence has been questioned by some, but it does exist and it forms a good basis for describing the perineum. Viewed from below, the perineal membrane is triangular (Fig. 10.3). It has a long free border at the back that stretches between the two anterior limits of the ischial tuberosities. The sides of the perineal membrane run along the ischiopubic rami and stop just short of the pubic symphysis. At this point there

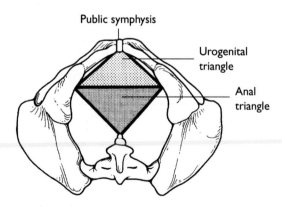

**Figure 10.1** Seen from below, a line drawn between the ischial tuberosities defines the urogenital triangle between itself and the pubic symphysis in front, and the anal triangle between itself and the tip of the coccyx posteriorly.

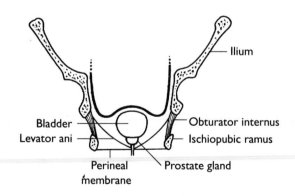

**Figure 10.2** The bladder and prostate lie above the pelvic diaphragm, which is convex to its lowest point in the midline. Peritoneum extends down to cover the upper aspect of the bladder and the other pelvic viscera. The perineal membrane lies beneath the pelvic diaphragm and runs between the ischiopubic rami.

is a gap between it and the **arcuate ligament** which joins the two pubic bones inferiorly.

The urogenital hiatus of the levator ani lies over the middle of the perineal membrane and close up against it. There is a hole through the middle of the perineal membrane for the urethra, and vagina in the female, to pass through. This is really just the same hole extended inferiorly as the urogenital hiatus containing the urethra (and vagina in the female). The fibrous perineal body that lies in the midline of the levator ani muscle is fused to the middle of the posterior margin of the perineal membrane. More laterally, the curved surface of levator ani sweeps upwards away from the perineal membrane creating an anterior recess for the ischiorectal fossa (Fig. 10.2) which we will study below.

Above the perineal membrane, between it and the undersurface of the levator ani, is a muscle called the **external sphincter urethrae**. There may be some small differences between the arrangement of the muscle fibres in males and females but, basically, the external sphincter urethrae probably arises in part from the bone of the inferior pubic ramus, runs towards the urethra and encircles it. The important point is that it encircles the urethra above the perineal membrane and is a voluntary sphincter in both sexes. Any other small muscles above the perineal membrane are both difficult to demonstrate and unimportant, and will not be described here since they distract from the importance of the external sphincter urethrae.

## The male perineum

Embedded in the sphincter urethrae in the male are the two **bulbourethral glands**. The ducts from these glands are quite long and first pierce the perineal membrane before opening into the urethra. Superficial to the perineal membrane in the male is the **root of the penis**. The root of the penis is formed by three masses of erectile tissue, each of which is surrounded

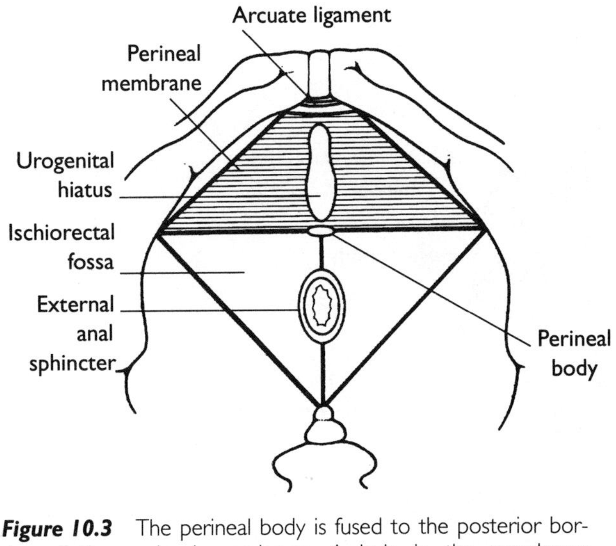

**Figure 10.3** The perineal body is fused to the posterior border of the perineal membrane. A hole in the membrane coincides with the urogenital hiatus in the pelvic diaphragm above. Anteriorly the membrane has a free edge just behind the pubic symphysis.

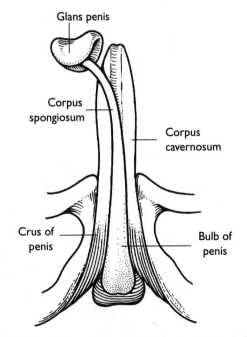

**Figure 10.4** The right and left crus of the penis with the bulb of the penis together form the root of the penis. This lies directly beneath the perineal membrane. The crura extend forwards into the free shaft of the penis as the corpora cavernosa and the bulb extends forwards as the corpus spongiosum. The glans penis is formed from the corpus spongiosum. (After Sabotta J and McMurrich JP. *Atlas of Human Anatomy.* New York: GE Stechert, 1930.)

by a tough fibrous sheath (Fig. 10.4). The **crura** of the penis are attached to the inner surface of each ischiopubic ramus. The crura pass forwards to meet each other at the anterior margin of the perineal membrane and continue into the free shaft of the penis. The crura become the two **corpora cavernosa** of the shaft. Each corpus cavernosus retains its separate tough fibrous covering within the shaft but there is free communication of blood vessels between the two through perforations in the fibrous septum along the length of the shaft.

The third component of the root of the penis is the **bulb of the penis**. It lies in the midline against the undersurface of the perineal membrane. The bulb of the penis gradually becomes cylindrical and then enters the free shaft of the penis anterior to the two corpora cavernosa. (Note that the anatomical position of the penis is erect.) Once in the free shaft of the penis the bulb of the penis is called the **corpus spongiosum**. The corpus spongiosum is longer than the two corpora cavernosa and extends beyond them to form the **glans penis** at the tip.

The bulb of the penis and the two crura are covered by muscles which, when they contract, may assist in initiating and maintaining erection of the penis. Take good note, however, that erectile tissue is a specialized tissue that does not function by obstruction of venous return. Erectile tissue contains 'corkscrew-like' or **helicine** arterioles that open into fibrous saccules or spaces. As a result of parasympathetic stimulation, the smooth muscle of the walls of the arterioles dilates and the saccules become taut and distended with arterial blood as the vessels enlarge. There is a huge difference between hot red erectile tissue and cold blue body parts where the venous return has become restricted.

Two **ischiocavernosus muscles** enclose the crura over the ischiopubic rami, and the **bulbospongiosus muscle** covers the bulb of the penis in the midline (Fig. 10.5). The bulbospongiosus muscle has a midline raphé or seam and has a sphincteric action that also assists in ejaculation and urination.

## The male urethra

The urethra in the male is long (Fig. 10.6). The **prostatic part of the urethra** is widest and most distensible and has a long ridge on its posterior wall. The ridge is called the **urethral crest**. In the summit of this crest is a small pit called the **prostatic utricle**. The prostatic utricle is all that remains in the male of the structures that develop into the vagina and uterus in the female.

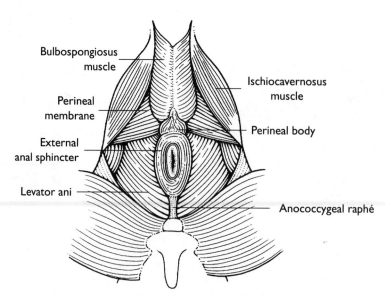

**Figure 10.5** The two ischiocavernosus muscles attach to the ischiopubic rami and cover each crus of the penis. The bulbospongiosus muscle lies around the bulb of the penis.

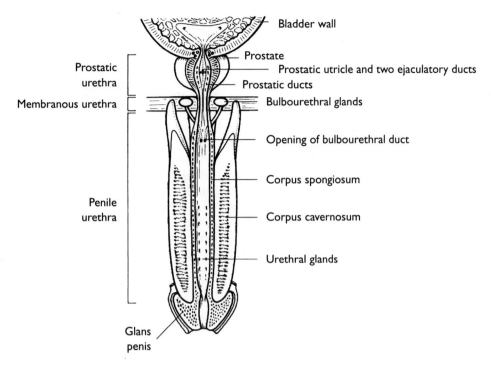

**Figure 10.6** The openings of the two ejaculatory ducts lie on either side of the prostatic utricle on the urethral crest of the prostatic urethra. The sphincter urethrae muscle lies around the membranous part of the urethra, immediately below the prostate gland and immediately above the perineal membrane. It has the bulbourethral glands embedded in it. The ducts of these glands pierce the perineal membrane and enter the proximal part of the penile urethra. The external urethral orifice is the narrowest part of the urethra. (After Green JH and Silver PHS. *An Introduction to Human Anatomy*. New York: Oxford Medical Publications, 1981.)

The two openings of the ejaculatory ducts lie on either side of the prostatic utricle. All the way along the urethral crest on either side there are additional small openings for the prostatic ducts of the prostate gland.

The part of the urethra surrounded on the outside by the external sphincter urethrae is much less distensible. It is called the **membranous part of the urethra**, not because its structure is any different but simply because it passes through the perineal membrane about here. The **penile urethra** pierces the bulb of the penis as it passes below the perineal membrane. It then travels forwards through the whole length of the shaft within the corpus spongiosum. Sometimes it is referred to as the spongy part of the urethra. The penile urethra opens at the **external orifice of the urethra**, which is the narrowest part of all. There are two expanded fossae within the penile urethra. One is the **navicular fossa** just distal to the external orifice and the other is the **intrabulbar fossa**. Along the length of the penile urethra there are openings for glands in the floor (or anterior wall), which create small folds.

## The female perineum

It is important to understand the homologies between the male and female genitalia. Look at Figure 10.7 and understand that early on in development at the 'indifferent stage' there is a **urethral groove** in the **phallus** which is flanked by **urogenital folds**. Paired **genital swellings** lie lateral to these. In males the urogenital folds and the genital swellings come together and fuse to form the shaft of the penis and the scrotum respectively. A raphé or seam marks this line of fusion. In females these structures remain separate. The urogenital folds and the genital swellings continue to grow and become the labia majora and labia minora. The urethra and vagina still open into what was the urethral groove.

In adults, the two crura of the clitoris lie on the ischiopubic rami and differ from those of the penis only in size (Fig. 10.8). They are of smaller diameter and the shaft of the clitoris is short. The two crura fuse anteriorly to form the **corpus clitoridis**. The **ischiocavernosus muscle** is smaller than the male counterpart but identical in every other way. There is no single bulb of the clitoris because the vagina and urethra open directly through the skin between the labia minora exactly where a bulb would lie. Within each labium minorum, though, is an erectile body called the **bulb of the vestibule**. The anterior pole of each bulb runs onto the anterior surface of the shaft of the clitoris. It then fuses with the bulb from the other side to form the **glans of the clitoris**. The bulbospongiosus muscle in the female is divided by the margins of the vagina. It passes posteriorly over each bulb of the vestibule and inserts into the fibrous perineal body behind the vagina. Its sphincteric action in the female is to narrow the vestibule of the vagina.

The urethra in the female is short. It runs from the internal urethral orifice in the bladder along the anterior wall of the vagina. It opens at the external urethral orifice in the vestibule between the opening

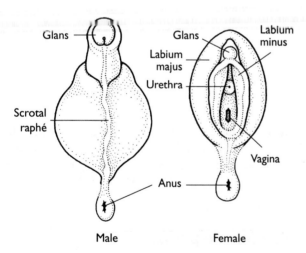

**Figure 10.7** At the indifferent stage in the early embryo, the genital swellings surround the urogenital folds which in turn lie on either side of the urethral groove. In the male the urogenital folds fuse and enclose the urethra, and the genital swellings also fuse across the midline to form the scrotum. In the female the urogenital folds remain unfused and become the labia minora which surround the openings of the urethra and vagina. The genital folds become the labia majora. (After Fitzgerald MJT and Fitzgerald M. *Human Embryology.* London: Baillière Tindall, 1994.)

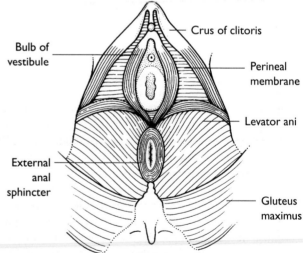

**Figure 10.8** The crura of the clitoris lie on the ischiopubic rami and are covered by the ischiocavernosus muscles. Anteriorly they fuse to form the body of the clitoris. The bulb of the vestibule is also fused anteriorly and forms the glans of the clitoris but is divided posteriorly and lies on either side of the urethral and vaginal openings within the labia minora. (After Snell RS. *Clinical Anatomy for Medical Students.* Boston: Little, Brown, 1981.)

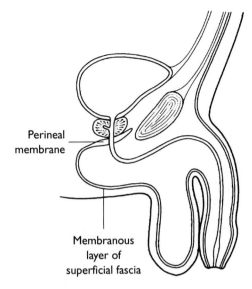

**Figure 10.9** The membranous layer of the superficial fascia attaches to the posterior border of the perineal membrane and runs anteriorly beneath the skin of the scrotum, around the root of the penis and up on to the anterior abdominal wall. (After Hall-Craggs ECB. *Anatomy as a Basis for Clinical Medicine*. Munich: Urban and Schwarzenberg, 1990.)

of the vagina and the clitoris. There is both an involuntary internal urethral sphincter at the internal urethral orifice of the bladder and a voluntary external sphincter urethrae immediately above the perineal membrane, exactly as there is in the male.

# Fascia covering the urogenital triangle

Because it is clinically significant, in the male at least, we need to consider the attachments of the superficial fascia overlying the structures that lie external to, or superficial to, the perineal membrane. The superficial fascia over the lower part of the abdomen and overlying the urogenital triangle consists of a deeper membranous portion and a superficial layer containing more loose connective tissue and fat. The membranous layer of superficial fascia attaches to the posterior free border of the perineal membrane (Fig. 10.9). It also attaches at the sides along the ischiopubic rami (Fig. 10.10). From here, the margin of attachment then runs up the front of the pubic bones and finally proceeds laterally over the front of the thigh. In this position the attachment of the membranous layer of superficial fascia fuses with the fascia lata of the thigh. From this fixed periphery the membranous layer runs forwards into the skin of the scrotum and shaft of the penis. Above, it is continuous with the membranous layer of superficial fascia of the anterior abdominal wall.

If urine and/or blood leaks from a ruptured urethra in the bulb of the penis or penile urethra it collects in what is called the **superficial perineal pouch**. (The concept of a **deep perineal pouch** above the perineal

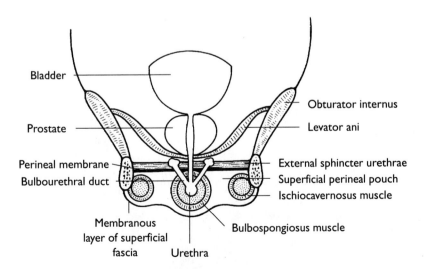

**Figure 10.10** The structures that form the root of the penis lie between the perineal membrane and the membranous layer of the superficial fascia. Damage to the urethra in the bulb of the penis here may result in extravasation of blood and urine into the superficial perineal pouch which can then only track forwards.

membrane is, we believe, undemonstrable and not worthy of description.) Urine or blood cannot pass up beyond the perineal membrane. It cannot pass laterally into the thigh or backwards into the anal triangle. But it may track forwards beneath the membranous layer of superficial fascia into the scrotum and penis, and eventually up into the anterior abdominal wall. Exactly how many patients ever present with this sort of discrete extravasation of urine from a ruptured urethra is uncertain but the concept of a superficial perineal pouch remains an important one.

# The anal triangle and ischiorectal fossa

We have already described the anal canal and the rectum in our study of the abdomen. The anal canal is the major structure within the anal triangle. The anal canal is surrounded by fatty connective tissue (Fig. 10.11). The space to the sides of the canal between it and the ischiopubic ramus is called the **ischiorectal fossa** (or, more accurately, the **ischioanal fossa**). The ischiorectal fossae are continuous with each other around the anal canal and with the space between the perineal membrane and the muscles of the pelvic diaphragm above, the **anterior recess** of the ischiorectal fossa. It extends up to the attachment of the arcus tendineus on the obturator internus.

# The pudendal nerve

The **pudendal nerve** is derived from sacral segments S2, S3 and S4. It is the nerve of the perineum. The nerve leaves the pelvic cavity through the greater sciatic foramen and then curls round the sacrospinous ligament to return through the lesser sciatic foramen. In this way the pudendal nerve finds its way into the perineum. Here, it runs in a canal in the fascia covering the inner aspect of the obturator internus muscle (Fig. 10.11). This is called the **pudendal canal** or **Alcock's canal**). At the point of entry into the fascia the nerve gives off the **inferior rectal nerve** which supplies the external anal sphincter and the skin around the anus. The pudendal nerve then runs forwards in the pudendal canal where it bifurcates into its two terminal branches, the **perineal nerve** and the **dorsal nerve of the penis** (or **dorsal nerve of the clitoris**). The perineal nerve travels forwards, superficial to the perineal membrane, giving off branches which supply the structures that lie on the superficial surface of the perineal membrane. It supplies the skin of the perineum and the skin of the posterior surface of the scrotum (or labia). The dorsal nerve of the penis (or clitoris) travels forwards on the superior, or deep, surface of the perineal membrane. It supplies the sphincter urethrae and any other muscle fibres that exist above the perineal membrane. At the anterior edge of the perineal membrane it runs into the posterior surface of the penis (or clitoris) and supplies the skin and fascia. The erectile components of the penis or clitoris are supplied by branches from the pelvic

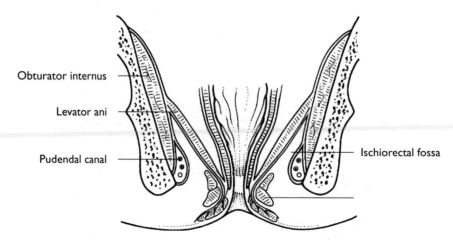

**Figure 10.11**    The space to the sides of the anal canal, between it and the ischiopubic ramus, is called the ischiorectal fossa.

Obturator internus

Levator ani

Pudendal canal

Ischiorectal fossa

plexus that pierce the pelvic diaphragm and perineal membrane.

# The internal pudendal artery and vein

For each branch of the pudendal nerve there is an accompanying branch of the **internal pudendal artery** (Figs 10.12 and 10.13). The internal pudendal artery continues in the pudendal canal above the perineal membrane, just medial to the ischiopubic ramus. It ends by dividing into the **deep** and **dorsal arteries of the penis**. It gives off the **inferior rectal artery** and the **perineal artery** before running above the perineal membrane. It gives three arteries to the penis. The **artery to the bulb of the penis** pierces the perineal membrane and enters the bulb to supply the erectile tissue of the corpus spongiosum. The **deep artery of the penis** supplies the erectile tissue of the corpus cavernosum. The **dorsal artery of the penis** runs between the deep fascia and fibrous sheath of the corpus cavernosum and supplies skin and superficial layers of the penis. Tributaries of the **internal pudendal vein** accompany the artery. One exception is the single **deep dorsal vein of the penis**. This drains to the prostatic venous plexus surrounding the prostate gland and not back to the internal pudendal vein.

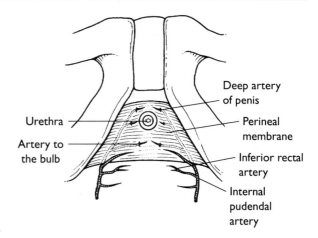

**Figure 10.13** The internal pudendal artery is distributed with the pudendal nerve. It divides into the dorsal nerve of the penis, running above the perineal membrane, and into the perineal branch which runs beneath it.

# The sacral plexus

The ventral rami of the 4th and 5th lumbar segments join together to form the **lumbosacral trunk** (Fig. 10.14). The lumbosacral trunk runs down into the pelvis to form the sacral plexus along with the first four sacral ventral rami. The sacral ventral rami leave the sacrum through the anterior sacral foramina. The sacral plexus then is made up of segments L4, L5, S1, S2, S3 and S4. The sacral plexus is a mixing of nerves and it lies on the surface of the piriformis muscle. Much of the sacral plexus is concerned with the nerve supply to the lower limb, and the **sciatic nerve** is by far the largest component. The sciatic nerve leaves the pelvis through the greater sciatic foramen. Other smaller nerves form in the sacral plexus and pass out, for example, to the gluteal region (**superior** and **inferior gluteal nerves**) and to the skin on the back of the thigh (**posterior femoral cutaneous nerve**). It is not surprising to find that piriformis gets its nerve supply from the sacral plexus.

The **pudendal nerve** forms in the sacral plexus from segments S2, S3 and S4. As we have seen this is a vitally important nerve to the perineum. The **nerves to levator ani** and **coccygeus** come directly from the 4th sacral ventral ramus.

The **pelvic splanchnic nerves** arise from the sacral plexus and are the parasympathetic outlet for the pelvic viscera and external genitalia as well as for the distal third of the transverse colon and descending

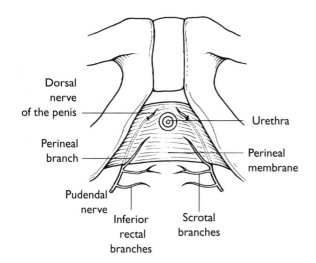

**Figure 10.12** The pudendal nerve supplies the structures of the perineum including the genitalia, as well as the external anal sphincter and anal mucous membrane below the level of the pectinate line.

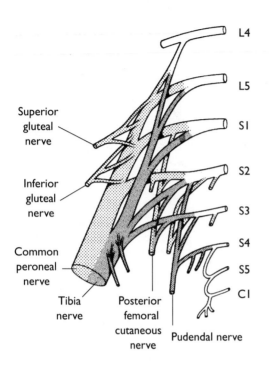

Superior
gluteal
nerve

Inferior
gluteal
nerve

Common
peroneal
nerve

Tibia
nerve

Posterior
femoral
cutaneous
nerve

Pudendal nerve

L4
L5
S1
S2
S3
S4
S5
C1

**Figure 10.14** The lumbosacral trunk and the sacral ventral rami join together to form the lumbosacral plexus. Anterior divisions of the ventral rami (dark shading) contribute to the tibial part of the sciatic nerve and posterior divisions of the ventral rami (light shading) to the common peroneal part. Parasympathetic pelvic splanchnic nerves arise from segments S2, S3 and S4. (After Hall-Craggs ECB. *Anatomy as a Basis for Clinical Medicine.* Munich: Urban and Schwarzenberg, 1990.)

colon. The pelvic splanchnic nerves arise from segments S2, S3 and S4.

# A summary of the autonomic nerves in the abdomen and pelvis

By now you will be aware that the autonomic nervous system is made up of parasympathetic and sympathetic components. Both the parasympathetic and sympathetic nerves comprise two axons which synapse in a ganglion somewhere between the spinal cord and the target organ being innervated. Parasympathetic nerve fibres leave the central nervous system only from the brain (in cranial nerves) or from sacral segments S2, S3 and S4 which arise in the sacral plexus. Sympathetic nerve fibres, on the other hand, leave the central nervous system only from the spinal cord between segments T1 to L2.

Sympathetic nerves leave the spinal cord and enter the sympathetic trunk (Fig. 10.15). The sympathetic trunk runs vertically either side of the vertebral column and extends up beyond the level of T1 to the base of the skull. It extends below the level of L2 and runs into the pelvis. The two sympathetic trunks unite over the coccyx at a small swelling known as the **ganglion impar**. In this way sympathetic nerve fibres can reach areas of the body either above or below the levels of their origin between T1 and L2. Sympathetic fibres from the sympathetic trunk in the pelvis run to join the sacral and coccygeal nerves and travel with them to the limbs and skin. A few are destined for pelvic organs and run to join the **inferior hypogastric plexuses** (see below).

## Sympathetic nerves in the abdomen and pelvis

Preganglionic sympathetic nerve fibres to all the abdominal and pelvic organs come from segments T5 to L2 of the spinal cord. They run out of the sympathetic trunk as **splanchnic sympathetic nerves**. In the thorax the greater, lesser and least splanchnic nerves arise from segments T5 to T12 and in the abdomen the **lumbar sympathetic splanchnic nerves arise from** segments T12 to L2. All the axons in the splanchnic nerves pass through the sympathetic trunk without synapsing. They synapse in ganglia (large or small) that lie along the anterior surface of the abdominal aorta. Collectively, they are referred to as **preaortic ganglia**. They are also given more specific regional names that often relate to the nearest major artery (Fig. 10.15). The **coeliac ganglia**, for example, are found around the coeliac trunk; **superior** and **inferior mesenteric ganglia** are found around the origin of the arteries of the same name. Below the inferior mesenteric artery there are small ganglia that extend along the front of the aorta and then on to the sacrum as far down as the second sacral segment (Fig. 10.16). These are known collectively as the **superior hypogastric ganglia**. Further down on either side of the rectum there are two groups of **inferior hypogastric ganglia**. Even lower down in the pelvis, within the subperitoneal space, there are yet other **pelvic sympathetic ganglia**.

Postganglionic sympathetic nerve fibres travel from the various preaortic ganglia along arteries to distribute to the the organs supplied by these vessels. Post-

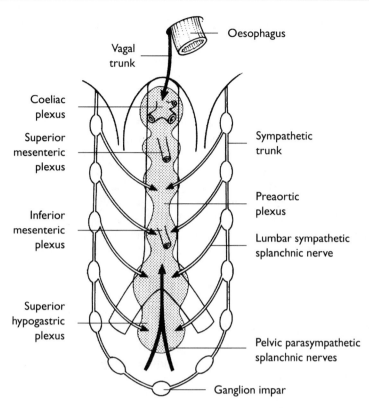

**Figure 10.15**   Preganglionic sympathetic splanchnic nerves pass to the preaortic plexuses from the sympathetic trunks. Preganglionic parasympathetic splanchnic nerves join the preaortic plexuses from the vagal trunks above and from sacral segments S2, S3 and S4 below. (After Hall-Craggs ECB. *Anatomy as a Basis for Clinical Medicine.* Munich: Urban and Schwarzenberg, 1990.)

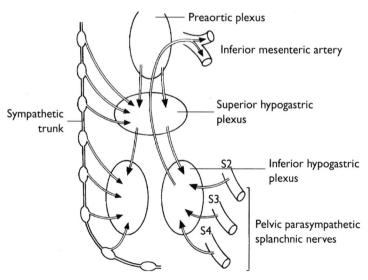

**Figure 10.16**   The superior hypogastric plexus and the inferior hypogastric plexuses are autonomic plexuses in the pelvis. They receive preganglionic sympathetic fibres from the sympathetic trunks and preaortic plexus, and preganglionic parasympathetic fibres from sacral segments S2, S3 and S4. Some of these parasympathetic fibres ascend to distribute with the inferior mesenteric artery. (After Hall-Craggs ECB. *Anatomy as a Basis for Clinical Medicine.* Munich: Urban and Schwarzenberg, 1990.)

ganglionic sympathetic nerve fibres from the superior and inferior hypogastric and pelvic ganglia, however, travel directly to pelvic organs.

### Parasympathetic nerves in the abdomen and pelvis

Preganglionic parasympathetic nerves to the foregut and midgut travel in the vagus nerve. The ganglia containing the synapses and postgangionic cell bodies and axons of the parasympathetic nerve fibres are found within the wall of the gut or within the target organs. Vagal preganglionic fibres mingle with the sympathetic nerve fibres. They run between the pre-aortic sympathetic ganglia and distribute to the viscera with them. It follows, for example, that the sympathetic coeliac ganglia and the vagal parasympathetic fibres that mingle with them in this region will be described together more generally as the **coeliac plexus** of nerves. This also follows for the other regional plexuses that lie in front of the aorta where there is a mixing of sympathetic and parasympathetic nerves.

The hindgut and the pelvic organs receive their preganglionic parasympathetic nerve fibres from the **pelvic parasympathetic splanchnic nerves** that arise within the sacral plexus. Understand that a **splanchnic nerve** is simply a nerve that runs to a viscus. It is better when trying to learn about them for the first time to describe them in full, for example pelvic *parasympathetic* splanchnic nerves or lumbar *sympathetic* nerves, for clarity. Just to confuse matters even more, certain pelvic parasympathetic splanchnic nerves are sometimes also known as the **nervi erigentes** (because they raise the penis; L erigo = to raise). Some of the preganglionic pelvic parasympathetic nerve fibres break 'the rule' we have just observed in the abdomen and synapse in tiny ganglia found in pelvic plexuses rather than in the walls of the organs they supply. Examples of these are the parasympathetic nerves that supply the erectile components of the penis and clitoris. These more than likely synapse in the tiny ganglia in pelvic plexuses and then run on as postganglionic nerve fibres which pierce the pelvic diaphragm and perineal membrane. From here they innervate the erectile tissue of crura and bulb of the penis or clitoris directly. Damage to the pelvic plexuses can occur in males during prostate operations since the plexuses lie around this gland. The result can be a problem in initiating or maintaining an erection. Most preganglionic parasympathetic

nerves pass through these pelvic plexuses and synapse in the walls of pelvic organs. Branches of pelvic parasympathetic splanchnic nerves, for example, travel directly upwards to the descending colon and rectum, and synapse in the wall of the gut.

It is thought that the parasympathetic system has exclusive control of the muscular wall of the bladder, urethra and rectum. The internal sphincters of the bladder and anal canal contract under sympathetic stimulation but are relaxed by parasympathetic activity. The parasympathetic pelvic splanchnics also cause vasodilatation of the arteries supplying the erectile tissue of the penis and clitoris. Combined with the effects of contraction of the ischiocavernosus and bulbospongiosus muscles, supplied by the perineal nerve, erection is initiated and maintained.

The sympathetic system in males causes the epididymis, vas deferens, seminal vesicles and prostate to contract and empty their contents. The sympathetic system is therefore involved in ejaculation. At the same time the neck of the bladder is closed off preventing a reflux of seminal fluid into the bladder. Sympathetic innervation of the uterus and ovaries is mainly vasomotor. Parasympathetic innervation of the cervix and vaginal wall is mainly to mucous glands.

## Visceral pain in the abdomen and pelvis

This is a difficult subject but one that helps explain many of the feelings of fullness, and emptiness of the 'stomach', bladder and bowel, for example, as well as pain from the viscera. Visceral sensation and pain from the abdominal and pelvic organs travel back to the spinal cord along with afferent autonomic nerves. These sensory nerve fibres follow exactly the same pathways as either the motor sympathetic or the motor parasympathetic pathways we have just described. In every other way they are exactly like other sensory neurons. They have one neuron with a cell body in a dorsal root ganglion.

Sensations other than pain from the foregut and midgut travel back through the vagus nerve along parasympathetic pathways. Pain from the abdominal viscera travels back in company with sympathetic pathways. Pelvic and perineal visceral pain travels back to the spinal cord mostly along pelvic parasympathetic pathways.

In general, pain returns to the dorsal roots of the segments of origin of the preganglionic sympathetic or parasympathetic nerve fibres. Pain is often then referred back along the same somatic segments. Early pain from the appendix, therefore, that is innervated by sympathetic fibres originating in segments T10 and T11 passes back to the dorsal roots of the spinal cord at T10 and T11 and is then referred back along the somatic distribution of nerves T10 and T11. It is felt at the farthest point from the spinal cord along this dermatome, which in this case is around the umbilicus. Pain from the gonads is felt around the umbilicus for exactly the same reason.

Pain from the uterine cervix, for example during childbirth, travels along the parasympathetic pelvic splanchnic pathways to the segments of origin, S3 and S4. Pain is then referred back along the sacral somatic distribution of these nerve segments. Pain from the uterine cervix is therefore felt mainly over the back of the sacrum which is the farthest point of this dermatome from the spinal cord. Pain from the body of the uterus travels in afferent fibres that follow the sympathetic nerves back to segments T12, L1 and L2. Pain from the body of the cervix will be felt, therefore, over the lower abdominal wall.

# chapter 11

# Revision of the Abdomen, Pelvis and Perineum

The following multiple choice questions are designed to help you consolidate your knowledge of the anatomy of the abdomen, pelvis and perineum. Many of these questions are quite searching and may require a knowledge of the peripheral nervous system, the relationships of various abdominal and pelvic structures as well as of their topography. Remember that any of the options (A)–(E) may be either correct or incorrect and that there is no pattern to the combination of correct answers. You may choose to do alternate questions on your first attempt and then on a subsequent occasion complete the remaining questions. You may choose to do them a few at a time in parallel with your progress through the text. You may simply prefer to attempt them all on one occasion and then repeat them all again when you have reread parts of the text. Whatever you choose, be sure to work through them in conjunction with the text. We expect you to have to use the text to discover the correct answers to some of the questions. By doing this you will improve your understanding of the anatomy of the abdomen, pelvis and perineum. At your first attempt, a score of about 50% correct responses would be quite reasonable. We would expect this to improve at repeated attempts of the same questions.

# Multiple Choice Questions on the Abdomen

## 1. The rectus sheath and/or rectus abdominis muscle:
(A) has four tendinous intersections
(B) has a fused midline seam called the arcuate line
(C) is incomplete anteriorly below the linea alba
(D) extends the vertebral column
(E) is supplied in part by the ilioinguinal nerve

A____ B____ C____ D____ E____

## 2. The inguinal ligament:
(A) is attached at one end to the anterior superior iliac spine
(B) is attached at the other end to the pubic symphysis
(C) is the lower, rolled, edge of the external oblique aponeurosis
(D) is sometimes pierced by the lateral cutaneous nerve of thigh
(E) is pierced by the ilioinguinal nerve

A____ B____ C____ D____ E____

## 3. The conjoint tendon:
(A) forms part of the anterior wall of the inguinal canal
(B) receives some of its fibres from the external oblique aponeurosis
(C) receives some of its fibres from the aponeurosis of transversus abdominis
(D) is found on the lateral side of the internal (deep) inguinal ring
(E) is attached to the pubic bone

A____ B____ C____ D____ E____

## 4. Structures passing through the deep inguinal ring in the male are:
(A) the inferior epigastric artery
(B) the ductus deferens
(C) the ilioinguinal nerve
(D) the testicular artery
(E) cremasteric muscle

A____ B____ C____ D____ E____

## 5. The ilioinguinal nerve:
(A) runs posterior to the kidney
(B) is entirely sensory
(C) enters the inguinal canal through the deep inguinal ring
(D) is a branch of the lumbar plexus
(E) supplies sensation to part of the scrotal skin

A____ B____ C____ D____ E____

## 6. The pampiniform plexus of veins:
(A) is found in the spermatic cord
(B) drains into the internal iliac veins on the left
(C) drains into the inferior vena cava on the right
(D) drains blood from the testes
(E) closely surrounds the testicular artery

A____ B____ C____ D____ E____

## 7. The spermatic cord contains:
(A) a branch of the abdominal aorta
(B) lymph vessels destined for the inguinal nodes
(C) a branch of the inferior epigastric artery
(D) sympathetic fibres
(E) the gubernaculum

A____ B____ C____ D____ E____

## 8. In the scrotum:
(A) parasympathetic fibres innervate the dartos muscle
(B) there is a membranous layer of superficial fascia
(C) the epididymis lies in front of the testis
(D) the left testis usually lies lower than the right
(E) the front of the testis is devoid of tunica vaginalis

A____ B____ C____ D____ E____

## 9. Psoas major:
(A) is attached to lumbar intervertebral discs
(B) is attached to the 12th rib
(C) flexes the hip joint
(D) is supplied by the lumbar plexus
(E) is covered by strong fascia

A____ B____ C____ D____ E____

## 10. The abdominal aorta:
(A) passes behind the median arcuate ligament of the diaphragm
(B) ends just below the level of the umbilicus
(C) gives inferior phrenic branches
(D) gives off lumbar arteries
(E) is directly related to the liver

A____ B____ C____ D____ E____

## 11. The inferior vena cava:
(A) ascends on the posterior abdominal wall on the right side of the aorta
(B) is entirely retroperitoneal in its abdominal course
(C) receives right and left suprarenal veins
(D) receives hepatic veins
(E) is related to the bare area of the liver

A____ B____ C____ D____ E____

## 12. Parietal peritoneum:
(A) covers the appendix
(B) has a nerve supply which is the same as that of the overlying muscles and skin
(C) has sensory fibres from the phrenic nerves where it covers the undersurface of the diaphragm
(D) lies in front of the kidneys
(E) forms the lesser omentum

A____ B____ C____ D____ E____

## 13. The lesser omentum:
(A) lies betweem the liver and stomach
(B) forms part of the anterior wall of the lesser sac
(C) contains the ligamentum teres in its free edge
(D) contains the common bile duct in its free edge
(E) is sometimes called the gastrosplenic ligament

A____ B____ C____ D____ E____

### 14. The left ureter:
(A) is lined with columnar epithelium
(B) is completely surrounded by peritoneum
(C) passes inferior to the uterine artery in the female
(D) receives some of its blood supply from vesical arteries
(E) lies on the left psoas muscle

A____ B ____ C ____ D ____ E ____

### 15. The right suprarenal gland:
(A) lies on the upper pole of the right kidney
(B) receives a blood supply from the right renal artery
(C) receives a blood supply directly from a branch of the aorta
(D) has a venous drainage by a single vein into the right renal vein
(E) receives preganglionic sympathetic fibres from the splanchnic nerves

A____ B ____ C ____ D ____ E ____

### 16. Anterior relationships of the left kidney include:
(A) the hepatorenal pouch
(B) the tail of the pancreas
(C) the spleen
(D) the iliohypogastric nerve
(E) the transversalis fascia

A____ B ____ C ____ D ____ E ____

### 17. The right kidney is related posteriorly to:
(A) the psoas major
(B) the quadratus lumborum
(C) the lateral arcuate ligament of diaphragm
(D) the subcostal nerve
(E) the sympathetic trunk

A____ B ____ C ____ D ____ E ____

### 18. Behind the left kidney is:
(A) the diaphragm
(B) the genitofemoral nerve
(C) the costodiaphragmatic recess of pleura
(D) the 12th rib
(E) the iliacus muscle

A____ B ____ C ____ D ____ E ____

### 19. Concerning the kidneys:
(A) both kidneys are completely retroperitoneal
(B) the left kidney is usually lower than the right
(C) the anterior surface of the left kidney is related to the pancreas
(D) the right kidney is related to the second (descending) part of the duodenum
(E) the left renal vein drains into the portal vein

A____ B ____ C ____ D ____ E ____

### 20. The common bile duct:
(A) is formed by the union of the right and left hepatic ducts
(B) lies to the left of the hepatic artery
(C) passes behind the first part of the duodenum
(D) opens into the second part of the duodenum
(E) has the gastroduodenal artery as a close relation in part of its course

A____ B ____ C ____ D ____ E ____

### 21. The jejunum:
(A) is thicker walled than the ileum
(B) contains plicae circulares
(C) contains Peyer's patches in its walls
(D) receives blood from the superior mesenteric artery
(E) receives four or five short arcades of blood vessels in its mesentery

A____ B ____ C ____ D ____ E ____

### 22. The third part of the duodenum:
(A) lies in front of the inferior vena cava
(B) lies in front of the superior mesenteric vessels
(C) is related to the head of the pancreas
(D) receives the accessory pancreatic duct
(E) is related to the inferior mesenteric vein

A____ B ____ C ____ D ____ E ____

### 23. The superior mesenteric artery:
(A) supplies no blood to the colon
(B) indirectly supplies the appendix
(C) has a middle colic branch
(D) is posterior to the first (superior) part of the duodenum
(E) is related to the uncinate process of the pancreas

A____ B ____ C ____ D ____ E ____

### 24. The gall bladder:
(A) concentrates bile about 100 times
(B) gains its arterial blood supply from a branch of the right hepatic artery
(C) has a fundus which lies at the level of the 9th right costal cartilage
(D) has a body that is related to the duodenum
(E) is completely devoid of peritoneal covering

A____ B ____ C ____ D ____ E ____

### 25. The spleen:
(A) lies deep to the 9th, 10th and 11th left ribs
(B) is separated by the diaphragm from the chest wall
(C) is related to the stomach
(D) has the tail of the pancreas close to its hilum
(E) is supplied by an artery, which is a branch of the superior mesenteric

A____ B ____ C ____ D ____ E ____

### 26. The pancreas:
(A) lies behind the lesser sac
(B) has a tail that is contained within the leaves of the gastrosplenic ligament
(C) has an uncinate process that lies in front of the superior mesenteric vessels
(D) has a head that lies in front of the common bile duct
(E) may have an accessory pancreatic duct

A____ B ____ C ____ D ____ E ____

## 27. The stomach:
(A) is supplied by arteries arising from the splenic artery
(B) is supplied by arteries arising from the coeliac trunk
(C) has a venous drainage passing equally to both portal and systemic venous systems
(D) is related to the body of the pancreas posteriorly
(E) is a retroperitoneal structure

A____ B ____ C ____ D ____ E ____

## 28. A portocaval anastomosis occurs between:
(A) azygos and left gastric veins
(B) epigastric veins and veins of the falciform ligament
(C) portal vein and left renal vein
(D) intercostal and spinal veins
(E) superior and inferior rectal veins

A____ B ____ C ____ D ____ E ____

## 29. Concerning embryological remnants found in the adult:
(A) the ligamentum arteriosum is closely related to the left recurrent laryngeal nerve
(B) the median umbilical ligament is the remains of an umbilical artery
(C) the ligamentum teres (of the liver) is the remains of an umbilical vein
(D) Meckel's diverticulum, when present, is found in the jejunum
(E) the fossa ovalis is found on the interatrial septum of the right atrium

A____ B ____ C ____ D ____ E ____

## 30. The portal vein:
(A) is formed behind the second part of the duodenum
(B) is formed by the union of the splenic and superior mesenteric veins
(C) divides into hepatic veins
(D) passes in front of the epiploic foramen (opening into lesser sac)
(E) receives blood from the lower end of the oesophagus

A____ B ____ C ____ D ____ E ____

## 31. The apppendix:
(A) arises from the anteroinferior aspect of the caecum
(B) has a mesentery
(C) has both transverse and longitudinal layers of muscle
(D) may lie behind the caecum (retrocaecal)
(E) is supplied by a branch of the posterior caecal artery

A____ B ____ C ____ D ____ E ____

## 32. The spleen:
(A) extends in front of the left anterior axillary line
(B) is a content of the lesser sac
(C) is related to the left colic flexure
(D) is normally palpable
(E) is related to pancreas

A____ B ____ C ____ D ____ E ____

# Answers to Multiple Choice Questions

| | | | | | | | | | | | | | | | | | | | |
|---|---|---|---|---|---|---|---|---|---|---|---|---|---|---|---|---|---|---|---|
| 1. | A F | B F | C F | D F | E F | 12. | A F | B T | C T | D T | E F | 23. | A F | B T | C T | D F | E T |
| 2. | A T | B F | C T | D T | E F | 13. | A T | B T | C F | D T | E F | 24. | A F | B T | C T | D T | E F |
| 3. | A F | B F | C T | D F | E T | 14. | A F | B F | C T | D T | E T | 25. | A T | B T | C T | D T | E F |
| 4. | A F | B T | C F | D T | E F | 15. | A T | B T | C T | D F | E T | 26. | A T | B F | C F | D T | E T |
| 5. | A T | B F | C F | D T | E T | 16. | A F | B T | C T | D F | E F | 27. | A T | B T | C F | D T | E F |
| 6. | A T | B F | C T | D T | E T | 17. | A T | B T | C T | D T | E F | 28. | A T | B T | C F | D F | E T |
| 7. | A T | B F | C T | D T | E F | 18. | A T | B F | C T | D T | E F | 29. | A T | B F | C T | D F | E T |
| 8. | A F | B T | C F | D T | E F | 19. | A T | B F | C T | D T | E F | 30. | A F | B T | C F | D T | E T |
| 9. | A T | B F | C T | D T | E T | 20. | A F | B F | C T | D T | E T | 31. | A F | B T | C T | D T | E T |
| 10. | A T | B T | C T | D T | E F | 21. | A T | B T | C F | D T | E F | 32. | A F | B F | C T | D F | E T |
| 11. | A T | B T | C F | D T | E T | 22. | A T | B F | C T | D F | E F | | | | | | |

# Multiple Choice Questions on the Pelvis and Perineum

## I. With respect to the bony pelvis:
(A) the triradiate cartilage fuses shortly after birth
(B) the sacroiliac joint is a secondary cartilaginous joint
(C) the sacrum is made up of four fused segments
(D) the iliopectineal line defines the brim of the true pelvis
(E) the obturator foramen is formed partly by the pubic bone and partly by the ischium

A ____ B ____ C ____ D ____ E ____

## 2. Regarding sex differences between the male and female pelvis:
(A) the female pelvic inlet is more heartshaped than that of the male
(B) the female superior pubic rami are relatively longer than those of the male
(C) the subpubic angle in the female is less than 90°
(D) the true pelvis in the male is taller than that of the female
(E) the anteroposterior diameter of the female pelvic outlet is greater than that of the male pelvis

A ____ B ____ C ____ D ____ E ____

## 3. The right pubic bone:
(A) is part of the right innominate bone
(B) is united with the pubis of the other side by means of a fibrous joint
(C) has a tubercle which gives attachment to the inguinal ligament
(D) fuses with ilium and ischium in the acetabulum soon after birth
(E) bears a pubic crest which marks the attachment of the pectineus muscle

A ____ B ____ C ____ D ____ E ____

## 4. Within the pelvis:
(A) the piriformis muscle leaves the pelvic cavity through the greater sciatic foramen
(B) the obturator internus muscle leaves the pelvic cavity through the greater sciatic foramen
(C) the sacral plexus of nerves lies on the anterior surface of the piriformis muscle
(D) the coccygeus muscle runs between the ischial spine and the coccyx
(E) the urogenital hiatus is present in both males and females

A ____ B ____ C ____ D ____ E ____

## 5. Levator ani:
(A) arises in part from the fascia overlying the obturator internus muscle
(B) converges to a midline raphe or seam posterior to the anal canal
(C) runs into the perineal body in front of the anal canal
(D) rises up into the pelvic cavity forming a convex dome
(E) is comprised in part of a 'sling' of muscle fibres that encircle the rectum

A ____ B ____ C ____ D ____ E ____

## 6. The urinary bladder:
(A) has a parasympathetic nerve supply from the vagus nerves
(B) is lined with transitional epithelium
(C) has a fibrous cord, the urachus, attached to its apex
(D) has the ureters entering at the superior angles of the trigone
(E) has a blood supply from branches of the internal iliac arteries

A ____ B ____ C ____ D ____ E ____

## 7. The rectum:
(A) receives arterial blood from branches of the inferior mesenteric artery
(B) contains pillars beneath the mucosa which contain arteries and veins
(C) is sensitive to pain and temperature above the level of the pectinate line
(D) is sensitive to stretch and distension above the level of the pectinate line
(E) is not covered by peritoneum over its lower one-third

A ____ B ____ C ____ D ____ E ____

## 8. The anal canal:
(A) lies below the level of the 'sling' formed by puborectalis
(B) is surrounded by an external anal sphincter composed of circular smooth muscle fibres
(C) is surrounded by an involuntary external anal sphincter supplied by the autonomic nervous system
(D) is supplied by sensory nerve fibres from the pudendal nerve below the level of the pectinate line
(E) has associated venous plexuses that form a portacaval anastomosis

A ____ B ____ C ____ D ____ E ____

## 9. The uterus:
(A) communicates with the vagina via the uterine tube
(B) the angle of anteversion is the angle between the cervical canal and the axis of the uterine cavity
(C) is anchored to the sacrum by the cardinal ligaments
(D) the round ligament of the uterus represents the remnant of the lowermost part of the gubernaculum
(E) the uterus receives a nerve supply from the pelvic plexuses

A ____ B ____ C ____ D ____ E ____

## 10. The prostate gland:
(A) lies anterior to the urethra
(B) is surrounded by a venous plexus that contains no valves
(C) can be palpated only *per rectum* when enlarged
(D) contains the common ejaculatory ducts within its substance
(E) is invaginated by a small depression from the urethral crest of the prostatic urethra that is homologous with the female uterus

A____ B ____ C ____ D ____ E ____

## 11. The pudendal nerve:
(A) is a branch of the sacral plexus
(B) leaves the pelvis through the greater sciatic foramen
(C) passes through the fat of the ischiorectal fossa
(D) gives an inferior rectal branch to supply the muscle of the ampulla of the rectum
(E) supplies the muscles of the urogenital triangle of the perineum

A____ B ____ C ____ D ____ E ____

## 12. Concerning pelvic viscera:
(A) the pudendal nerve is formed from pelvic splanchnics
(B) ejaculation is under parasympathetic control
(C) the narrowest part of the urethra is the membranous urethra
(D) erectile tissue is under parasympathetic control
(E) the sphincter urethrae is a flat sheet of muscle which forms part of the pelvic diaphragm

A____ B ____ C ____ D ____ E ____

## 13. Concerning pelvic viscera:
(A) the trigone of the bladder is insensitive to pain
(B) the posterior fornix is related to the rectouterine pouch
(C) infection of the testis may give rise to enlarged inguinal lymph nodes
(D) the pampiniform plexus drains into the prostatic venous plexus
(E) The deep dorsal vein of the penis is single, the arteries of the penis are paired

A____ B ____ C ____ D ____ E ____

## 14. The right ovary:
(A) has a blood supply from the abdominal aorta
(B) has the ovarian ligament attached to its lateral pole
(C) is drained by veins which go to the inferior vena cava on the left
(D) is attached to the anterior (inferior) layer of the broad ligament
(E) has a lymphatic drainage to the inguinal nodes

A____ B ____ C ____ D ____ E ____

## 15. Structures opening directly into the prostatic part of the urethra include:
(A) the bulbourethral glands
(B) the ejaculatory ducts
(C) the seminal vesicles
(D) the ductus deferens
(E) the greater vestibular glands

A____ B ____ C ____ D ____ E ____

## 16. The penis:
(A) has a corpus spongiosum which ends posteriorly in the superficial perineal pouch as the bulb of the penis
(B) has corpora cavernosa which receive blood from the internal pudendal artery
(C) has a glans from which lymph drains into the para-aortic nodes
(D) has a venous drainage to the prostatic venous plexus
(E) is the homologue of the clitoris in the female

A____ B ____ C ____ D ____ E ____

## 17. The uterine artery:
(A) arises directly from the abdominal aorta
(B) runs to the uterus in the base of the broad ligament
(C) anastomoses with the ovarian artery of the same side
(D) crosses below the ureter as it approaches the uterus
(E) supplies the ovary

A____ B ____ C ____ D ____ E ____

## 18. The rectum:
(A) has a mesentery
(B) lies in front of the sacrum
(C) is directly related to the female bladder
(D) receives some arterial blood from a branch of the aorta
(E) possesses both circular and longitudinal smooth muscle fibres

A____ B ____ C ____ D ____ E ____

## 19. The vagina:
(A) has an anterior wall covered by peritoneum
(B) has the urethra attached to its anterior wall
(C) has lateral fornices related to the ureters
(D) is situated entirely below the pelvic diaphragm
(E) is separated from the anal canal by the perineal body

A____ B ____ C ____ D ____ E ____

## 20. Autonomic plexuses in the abdomen and pelvis:
(A) receive input from the genitofemoral nerve
(B) are found around the aorta and its branches
(C) receive an input from S2 S3 and S4 spinal nerves
(D) receive an input from the lumbar splanchnic nerves
(E) receive an input from the vagi

A____ B ____ C ____ D ____ E ____

# Answers to Multiple Choice Questions

| | | | | | | | | | | | | |
|---|---|---|---|---|---|---|---|---|---|---|---|---|
| 1. | AF | BF | CF | DT | ET | 8. | AT | BF | CF | DT | ET | 15. AF BT CF DF EF |
| 2. | AF | BT | CF | DT | ET | 9. | AF | BF | CF | DT | ET | 16. AT BT CF DT ET |
| 3. | AT | BF | CT | DF | EF | 10. | AF | BT | CF | DT | ET | 17. AF BT CT DF EF |
| 4. | AT | BF | CT | DT | ET | 11. | AT | BT | CF | DF | ET | 18. AF BT CF DT ET |
| 5. | AT | BT | CT | DF | ET | 12. | AF | BF | CF | DT | EF | 19. AF BT CT DF ET |
| 6. | AF | BT | CT | DT | ET | 13. | AF | BT | CF | DF | ET | 20. AF BT CT DT EF |
| 7. | AT | BT | CF | DT | ET | 14. | AT | BF | CF | DF | EF | |

# Index

Numbers in *italics* refer to illustrations